Collins *gem*

Butterflies

D0726625

Michael Chinery

HarperCollins*Publishers*
Westerhill Road, Bishopbriggs, Glasgow G64 2QT

www.collins.co.uk

First published 1995
This edition published 2004

10 09 08 07 06

10 9 8 7 6 5

The copyright in the photographs belongs to the following photographers from the Frank Lane Picture Agency:
Ted Benton 16, 17, 18, 21, 23, 26, 27, 33, 34, 36, 37, 40, 48, 49, 51, 53, 56, 57, 59, 60, 61, 62, 64, 65, 66, 67, 69, 70, 72, 73, 74, 79, 83, 84, 86, 89, 91, 92, 94, 95, 96, 98, 99, 100, 101, 104, 105, 106, 107, 111, 112, 113, 117, 118, 121, 122, 123, 127, 128; R. Bird 88; Borrell 212; H. D. Brandl 24, 177; M. Chinery 14, 32, 45, 55, 93, 129, 133, 135, 141, 148, 167, 172, 178, 181, 185, 187, 199, 200, 206, 209, 210, 217, 224, 227, 228, 231, 243, 244, 245; H. Clark 174; E. A. Dean 151; G. Dickson 201, 236; M. Evans 131, 147, 156, 160, 195, 208, 220, 221, 222, 232, 235, 237, 249; Ferrari 44, 47; A. R. Hamblin 251; Harmer 31, 90; Heard 54; E. & D. Hosking 144, 188, 216; J. Hutchings 218; G. E. Hyde 124, 140, 146, 149, 152, 154, 155, 157, 164, 165, 170, 182, 191, 193, 197, 198, 202, 203, 211, 213, 214, 225, 229, 230, 233, 234, 239, 240, 246, 247, 248; P. H. Jerrold 150, 215, 223, 226; Kreutzer 28; Ojalainen 29, 30; Panda Photo 136, 137, 176; G. Perrone/Panda 132; A. J. Roberts 205; I. Rose 19, 39, 52, 103, 145, 162, 163; M. Rose 186; Silvestris 143, 173, 175, 183, 184, 204, 250; J. Swale 219; J. Tinning 77, 179; A. Wharton 20, 102, 138, 139, 142, 153, 166, 168, 171, 189, 190, 192, 207, 238, 241, 242; R. Wilmshurst 50, 108, 134, 158, 196; M. B. Withers 159, 161, 169, 180, 252; B. Yates-Smith 13, 15, 22, 25, 35, 38, 41, 42, 43, 46, 58, 63, 68, 71, 75, 76, 78, 80, 81, 85, 87, 97, 109, 110, 114, 115, 116, 119, 120, 125, 126, 130.

Illustrations from *Collins New Generation Guide Butterflies and Moths of Britain and Europe* and *Collins Field Guide Caterpillars*, artist Brian Hargreaves

ISBN-13 978 0 00 717852 0
ISBN-10 0 00 717852 2

Design by Liz Bourne
Printed in Italy by Amadeus

Contents key

The insects described in this book belong to 22 families. Each of these families is briefly described on the next few pages, and each one has a small symbol showing something of the characteristic shape or colour of its members. Not all members exhibit the typical features of the family but, by comparing the symbols with your specimens, you should be able to place many of your finds in their correct families and then turn to the right place in the book. Conversely, having found a species in the main body of the book, you can use the symbol there to find the family description in this key for more information.

BUTTERFLIES
Skipper Family (*Hesperiidae*) pp. 13–21
Small brown or greyish butterflies with rapid, darting flight and rather plump, moth-like bodies. The antennae are curved or hooked at the tip. Many bask with their forewings partly raised.

Swallowtail Family (*Papilionidae*)
pp. 22–26
Large, brightly coloured butterflies, many of which have a little 'tail' on the hindwing. The inner margin of the hindwing is concave, leaving a clear gap between the body and the wings.

White Family (*Pieridae*) pp. 27–40
The European members of this family are all basically white or yellow, often with marked differences between the sexes.

Fritillary Family (*Nymphalidae*) pp. 41–72
A large and varied family whose members have only four walking legs. The front legs are small and brush-like and the insects are commonly called brush-footed butterflies. Most of them are brightly coloured, with reds and oranges being the dominant colours.

Brown Family (Satyridae) pp. 73–98
These butterflies are predominantly brown and almost all have eye-spots on the wing margins. They have only four walking legs. Their caterpillars nearly all feed on grasses.

Nettle-tree Butterfly Family (Libytheidae) p. 99
There is only one European member of this family, which is characterized by very long palps.

Duke of Burgundy Family (Riodinidae) p. 100
There is only one European member of this family.

Blue Family (Lycaenidae) pp. 101–130
A very large family of rather small butterflies, many of which exhibit gleaming metallic colours. Most European species are blue, although the females are often brown, but the family also includes the coppers, with their brilliant orange uppersides, and the hairstreaks named for the slender stripes on their undersides. Many have little 'tails' on the hindwings.

MOTHS

Swift Moth Family (Hepialidae) p. 131
Forewings and hindwings in this family are all the same shape and are folded along the sides of the body at rest. The antennae are very short and the moths have no tongue.

Goat Moth Family (Cossidae) pp. 132–133
Sturdy moths with wings folded back along the sides of the body at rest. They have no tongue and do not feed.

Burnet Family (Zygaenidae) *pp. 134–138*
Brightly coloured day-flying moths with long, clubbed antennae. Most are either black and red (burnets) or metallic green (foresters). The burnets' colours warn birds that they are poisonous – some of them actually contain cyanide!

Geometer Family (Geometridae) *pp.139–164*
A large family of mostly rather flimsy moths that generally rest with their wings spread flat on each side of the body. Most have rather pale or sombre colours. Their caterpillars alternately loop and stretch their bodies as they walk and are commonly called loopers.

Hooktip Family (Drepanidae) *p. 165*
A small family, named for the hooked wing-tips of most of the species. At rest, the wings are either folded along the sides of the body or spread flat like those of the geometers.

Lutestring Family (Thyatiridae) *pp. 166–167*
A small family, named for the fine lines that cross the wings of some species. The wings are held roofwise at rest, with little or no overlap.

Eggar Family (Lasiocampidae) *pp. 168–172*
Sturdy moths, covered with dense hair and mostly brown or yellow with little pattern. The tongue is short or absent and the moths do not feed. The wings are usually held roofwise over the body at rest.

Emperor Family (Saturniidae) *pp. 173–176*
Large, furry moths with an eye-spot on each wing. The male antennae are very feathery and can pick up the females' scents and home in on them from several kilometres. The tongue is absent.

Hawkmoth Family (Sphingidae) *pp. 177–191*
Large and generally fast-flying with narrow, pointed forewings. At rest, the wings are usually laid flat over the body and swept back into an arrowhead shape. The caterpillars generally have a small horn at the rear.

Prominent Family (Notodontidae) *pp. 192–200* A family of mostly sombre-coloured moths, whose wings are generally held steeply roofwise or wrapped tightly around the body at rest. Many have a tuft of scales on the rear of the forewing, and this forms a prominent hump when the moths are at rest.

Tussock Family (Lymantriidae) *pp. 201–205* Very hairy moths, rarely with much colour, whose wings are generally held roofwise at rest with the hairy legs prominently displayed. The tongue is very short or absent.

Syntomid Family (Ctenuchidae) *p. 206* Day-flying moths with weak, drifting flight. There are just a few species in southern Europe, all either black or brown with white spots.

Tiger Moth Family (Arctiidae) *pp. 207–219* The tiger moths are mostly large, brightly coloured and rather hairy moths, although most of the footmen are small and rather drab. Tiger moths rest with their wings held roofwise, but most footmen lay their wings flat with a good deal of overlap.

Noctuid Family (Noctuidae) *pp. 220–252* The largest of all moth families, mostly with sombre forewings although the hindwings may be brightly coloured. Most species rest with their wings held roofwise or else laid flat over the body with a considerable overlap.

The butterflies and moths belong to one of the largest groups of insects, with over 160,000 known species. More than 5,000 species occur in Europe, but most of these are very small and inconspicuous moths, as little as 3 mm across the wings and commonly known as 'micros'. The insects described and illustrated in this book are just a small selection of the European species, but they include many of the common and conspicuous species that are likely to come to your notice. Almost all the British butterflies are included. The photographs show the insects in their natural attitudes, which is how you normally see them in the field, so identification should be relatively easy.

The group is technically known as the Lepidoptera, a name meaning 'scale-wings' and referring to the minute scales that clothe the wings and give them their colourful patterns. The scales are only loosely attached and readily come off when the insects are handled. Underneath the scales, the wings have networks of veins that are used in classifying the insects into families. The cell is a more or less oval area near the front of the wing and its pattern is sometimes of use in separating closely related species of butterflies.

BUTTERFLY OR MOTH?

The division of the Lepidoptera into butterflies and moths is a very unequal one, for there are less than 20,000 butterfly species and they account for only about 15 of the 80 or so families in the order. There is no single difference between all the butterflies on the one hand and all the moths on the other, but the antennae or feelers are a pretty good guide: all butterflies have little knobs or clubs at the ends of their antennae, but very few moths have them. Among the European moths, only the burnets (pp.134–138) have clubbed antennae, and these can be recognized quite easily by their general shape and colouring. Almost all butterflies rest with their wings closed vertically over their bodies so that only the undersides of the wings are visible, whereas most of our moths rest with their wings spread flat or folded over the body with their uppersides visible.

FEEDING

Butterflies and moths feed on liquids, which they suck up through a tubular tongue or proboscis. This is coiled up under the head when not in use. Nectar from flowers is the major food of both butterflies and moths, but some species also enjoy fruit juices and some woodland butterflies, including the Purple Emperor and the Speckled Wood, exist largely on the sugary honeydew dropped by aphids. A number of moths do not feed in the adult state and their tongues are poorly developed or absent altogether.

THE LIFE HISTORY

Male and female moths usually find each other by means of scent. It is usually the female that releases the scent, and the male tracks her down with the aid of his antennae. These are often rather feathery with a large surface area with which to pick up the scent. Among the butterflies, however, the initial attraction is by sight, and only when the butterflies get close does scent, produced by both partners, come into play. The male scents are released mainly from special scales on the wings. These scales, known as androconia, are often clustered in distinct patches called scent brands.

The females lay eggs soon after mating, usually taking care to lay them on the correct foodplants for the resulting caterpillars. The eggs usually hatch within a couple of weeks, although a number of species pass the winter in the egg stage.

The caterpillars, or larvae, hatching from the eggs all have three pairs of true legs at the front corresponding to the legs of the adults and most have five pairs of stumpy legs called prolegs further back. The prolegs on the last segment are called claspers, and you'll see why if you try to remove a caterpillar from a twig! Some groups of caterpillars have even fewer prolegs, the best known being the loopers or geometer larvae, which have only two pairs including the claspers. Caterpillars have biting jaws instead of the sucking probosci of the adults. Most are leaf-eaters, but there are also caterpillars that eat roots and fruits and some that even chew their way through woody stems.

The young caterpillar embarks on an almost non-stop orgy of eating and very soon outgrows its tough skin, which does not grow with the rest of the body. It then takes a short rest and bursts out of its skin, having already grown a looser one beneath it. As soon as the new skin has hardened the caterpillar starts eating again, and a second moult is soon necessary. Most caterpillars undergo four such moults during their lives, and may be fully grown in under four weeks although caterpillars hatching in late summer often go into hibernation and complete their growth in the spring.

When fully grown, the caterpillar prepares to turn into a chrysalis, or pupa. This is a non-feeding stage, during which the larval body is broken down and converted into the adult form. Many moth caterpillars burrow into the ground and pupate in silk-lined chambers, while others spin silken cocoons around themselves before turning into pupae on or near their foodplants. The pupae are usually bullet-shaped and most are brown and shiny. Some butterfly caterpillars spin flimsy cocoons on the ground, but most butterfly pupae are naked. Some hang upside down from the foodplant or other support, while others are attached in an upright position and supported by a silken girdle. Being freely exposed in this way, the butterfly pupae need some kind of camouflage and are often ornately shaped and coloured to blend in with their surroundings.

When the adult insect, or imago, is ready to emerge, it bursts through the pupal skin and drags itself out. The wings are small and crumpled at first, but blood is then pumped into the veins and the wings soon expand and harden, and the insect is then ready to fly away.

Many butterflies and moths produce just a single generation or brood each year, the adults flying for just a few weeks at the appropriate season and then disappearing until the following year. Their offspring pass the winter as resting eggs, caterpillars, or pupae, according to the species. A few species, such as the Brimstone and Peacock butterflies, pass the winter as sleeping adults, and here the adult stage can last for almost a year.

Many other species have two broods in a year, with adults flying in late spring and again in the summer. The caterpillars hatching from eggs laid in the spring grow up rapidly, but those hatching in the summer may not produce adults until about nine months later.

A species that is single-brooded in the northern parts of its range may have two or even three broods further south. Upland populations also tend to be single-brooded, and those species living in the coldest regions may even take two years to complete their life cycles. Some wood-feeding and root-feeding caterpillars also take more than one year to grow up.

SPECIES ENTRIES

English and scientific names are given for most of the species described in the book, but a few non-British moths have never received English names and only the scientific name can be given. The sexes are mentioned in the descriptions only if there are significant differences between males and females.

ILLUSTRATION SYMBOLS

The illustrations of the butterflies in the book show both upperside and underside (undersides marked with ▲), male (♂) and female (♀) markings. Where no symbol is shown, the markings of both sexes are similar.

SIZE is the average length of a single forewing, measured from the shoulder to the wing-tip, although ranges are given for species that are particularly variable or in which males and females differ significantly in size.

HABITAT gives the main type of countryside in which the insect is likely to be found. Many species are restricted to areas where their foodplants grow.

FOODPLANT is the plant on which the caterpillars feed.

RANGE is given in a simplified form. N Europe covers Scandinavia and Denmark and is essentially that region above 55° N, although not including Scotland. The far N is the Arctic region, beyond the Arctic Circle. C Europe is Central Europe, roughly between 45° and 55° N but including all of the British Isles. S Europe is everything south of 45° N, from Bordeaux, through Turin, to the Danube Delta. Remember that the insects do not occur everywhere within their ranges – only where there is suitable habitat.

FLIGHT gives the flight times throughout the range of the insect, but these times may be much more restricted in northern areas, where adults generally appear much later.

SIMILAR SPECIES are those with which the insect can be confused: the text points out the differences.

caterpillar

An inconspicuous insect, making short, low-level flights in the sunshine. The upperside is deep brown with yellow patches of varying size and shade, usually paler in the female. The underside is brownish with a variable dusting of yellow scales and pale yellow spots on the hindwing. Winter is passed as a caterpillar.

SIZE 14 mm.

HABITAT Light woodland.

FOODPLANT Purple moor grass and other grasses.

RANGE Most of Europe except Mediterranean; confined to W Scotland in Britain.

FLIGHT June–July.

SIMILAR SPECIES Duke of Burgundy (p.100) has similar flight, but has black spots on wing margins. Northern Chequered Skipper is much paler.

Large Chequered Skipper

Heteropterus morpheus

♀

The yellow and white pattern of the underside of the hindwing is quite unmistakable. The upperside, rarely seen at rest, is dark brown with just a few small pale dots on the forewing. The butterfly dances over the vegetation in a jerky fashion, almost like a puppet on a string. Winter is passed as a caterpillar.

♂

SIZE 18 mm.

HABITAT Woodland clearings and damp grassland.

FOODPLANT Purple moor grass and other grasses.

RANGE Mainly E Europe; also N Spain and W France. Very local; threatened in many areas by drainage of wetlands.

FLIGHT June–August.

SIMILAR SPECIES None.

Both sexes have an orange upperside, but the female forewing lacks the black streak, containing the scent scales. The underside is dull orange tinged with grey. The antenna tip is orange below. The insect basks with its forewings partly raised. Winter is passed as a caterpillar.

SIZE 14 mm.

HABITAT Grassland, including woodland rides.

FOODPLANT Yorkshire fog and other tall grasses.

RANGE S & C Europe, except Scotland and Ireland.

FLIGHT June–August.

SIMILAR SPECIES Essex Skipper is almost identical, but has black-tipped antennae.

caterpillar

Look for the faint golden ring on the forewing – more obvious in the female (above right), than in the more orange male (above left), which also has a prominent black streak. The underside is dull orange in both sexes and the antenna tip is cream below. The insect basks with its forewings partly raised. Winter is passed as a caterpillar.

SIZE 12 mm.

HABITAT Open, grassy places.

FOODPLANT Tor grass.

RANGE S coast of England and most of S & C Europe.

FLIGHT May–September.

SIMILAR SPECIES Small Skipper (p.15) resembles male, but is greyer below, with orange antenna tips.

The upperside is rich brown with orange patches in both sexes, with a conspicuous black scent streak on the male's forewing. The underside is greenish with large silver spots. The insect basks with its forewings partly raised. Winter is passed as an egg. This is one of Britain's rarest butterflies, restricted to a few chalk and limestone hillsides in southern England.

SIZE 15 mm.

HABITAT Rough grassland with areas of bare soil.

FOODPLANT Sheep's fescue and other fine-leaved grasses.

RANGE Most of Europe.

FLIGHT June–September.

SIMILAR SPECIES Large Skipper (p.18) has similar upperside, but no silver spots on underside.

The upperside is rich brown with orange patches in both sexes, but the male has a conspicuous black scent streak. The underside is largely orange-green with faint yellow patches. It commonly perches on shrubs and basks with its forewings partly raised. Winter is passed as a caterpillar.

SIZE 16 mm.

HABITAT Rough grassland with shrubs; especially fond of woodland rides and hedgerows.

FOODPLANT Cocksfoot and other tall grasses. ♀

RANGE Almost all Europe except Ireland.

FLIGHT May–September, with 1–3 broods, but single-brooded in Britain and other northern areas.

SIMILAR SPECIES Silver-spotted Skipper (p.17) has greener underside with conspicuous silver spots.

The grey pattern on the forewing of this very moth-like butterfly varies in intensity. Look for the small white dots around the edges. The underside is pale brown with indistinct white dots. Flight is swift and low. The insect commonly perches with wings wide open, but wraps the wings around the body at night. Winter is passed as a caterpillar.

SIZE 14 mm.

HABITAT Open grassland.

FOODPLANT Birdsfoot trefoil and other legumes.

RANGE Most of Europe except far north. The only skipper in Ireland.

FLIGHT April–August, in one or two broods.

SIMILAR SPECIES Inky Skipper of SE Europe has similar upperside, but is dark brown below.

caterpillar

A moth-like butterfly whose low, darting flight is difficult to follow. The pattern varies in intensity, but the white spots are usually very clear and there is a large, more or less square one in the centre of the hindwing. The underside is a paler version of the upperside. The butterfly usually basks with wings wide open, often on the ground. Winter is passed as a pupa.

SIZE 12 mm.

HABITAT Rough grassland, heaths and woodland rides.

FOODPLANT Wild strawberry, cinquefoils and other low-growing plants, including trailing brambles.

RANGE Most of Europe, except Scotland and Ireland.

FLIGHT April–August, in one or two broods.

SIMILAR SPECIES None in Britain; several on the Continent, mostly with fewer markings on hindwing.

caterpillar

Named for the two conspicuous yellow bands on the underside of the hindwing. The upperside is brownish black with grey hairs, heavily marked with white, especially on the forewing. The hindwing has a clear row of white spots near the outer margin. The butterfly perches with wings open or closed. Winter is passed as a caterpillar.

SIZE 18 mm.

HABITAT Open, flower-rich grassland.

FOODPLANT Mallows.

RANGE S Europe except Iberia.

FLIGHT May–July.

SIMILAR SPECIES
Several skippers are similar above,
but no other has yellow bands on underside.

Apollo

Parnassius apollo

A rather furry butterfly whose thinly-scaled wings have much the same pattern on both surfaces. The spots vary a lot in size and may be red, orange or yellow. Flight is slow and rather laboured and it seems reluctant to fly. Winter is passed as an egg or caterpillar. The butterfly is becoming rare and is legally protected in several countries.

SIZE 40 mm.

HABITAT Mountain slopes.

FOODPLANT Stonecrops and houseleeks.

RANGE Most European mountains, except British Isles and far N.

FLIGHT June–August.

SIMILAR SPECIES Small Apollo is a little smaller, with two red spots on forewing.

caterpillar

The wings are essentially yellow with black and red markings, but look for the clear 'window' near the tip of the forewing. The underside is a paler version of the upperside, but with a white band on the hindwing. Winter is usually passed as a pupa.

SIZE 22 mm.

HABITAT Rough, stony hillsides and roadsides.

FOODPLANT Birthworts.

caterpillar

RANGE Iberia and S France.

FLIGHT Usually February–June, but double-brooded in far S where it may fly all year except July & August.

SIMILAR SPECIES Southern Festoon of SE Europe, including SE France, has no 'window' on forewing.

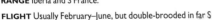

Common Swallowtail

Papilio machaon

A large, restless butterfly, rarely sitting still for long and often beating its wings while feeding at flowers. Upper and lower surfaces are alike, although the underside is slightly paler. Summer broods are paler than spring ones. Winter is passed as a pupa.

SIZE 35 mm.

HABITAT Grassland and other flowery places.

FOODPLANT Fennel, milk parsley and other umbellifers.

RANGE All Europe, but confined to a few East Anglian fens in Britain.

FLIGHT April–September, in 1–3 broods.

SIMILAR SPECIES None in Britain. Very rare Corsican Swallowtail has shorter 'tails' and very wavy black band on forewing.

caterpillar

A restless, fast-flying butterfly with conspicuous black stripes on a bright yellow background and a continuous blue band on the hindwing. The forewing is yellow at the base. The two surfaces are almost identical. Winter is passed as a pupa.

caterpillar

SIZE 32 mm.

HABITAT Flowery mountainsides; very fond of thistles.

FOODPLANT Fennel and other umbellifers.

RANGE S Europe from France eastwards, very local.

FLIGHT April–July.

SIMILAR SPECIES Common Swallowtail (opposite) has black base to forewing. Scarce Swallowtail (p.26) is paler, with longer 'tails'.

Scarce Swallowtail

Iphiclides podalirius

Upperside and underside are almost identical, with long 'tails' and black stripes on a cream or white background. Flight is fast, with long, elegant glides. Winter is passed as a pupa. Although a scarce visitor to Britain, this species is actually more common than the Common Swallowtail in most parts of Europe.

caterpillar

SIZE 40 mm.

HABITAT Open country and light woodland, including gardens and orchards.

FOODPLANT Blackthorn and various orchard trees.

RANGE S & C Europe.

FLIGHT March–September, in two broods.

SIMILAR SPECIES Southern Swallowtail (p.25) is brighter yellow with shorter 'tails' and continuous blue band on hindwing.

A flimsy butterfly with a weak fluttering flight, often seen dancing up and down in one spot for long periods. Its wings, more oval than those of the other whites, are never opened at rest. The male's forewing has a dark tip on its upperside, but the female wings are almost unmarked. The antenna has a white spot under the tip. Winter is passed as a pupa.

SIZE 22 mm.

HABITAT Woodland rides and lightly wooded country.

FOODPLANT Meadow vetchling and some other legumes.

RANGE Most of Europe except N Britain.

FLIGHT April–September, with 1–3 broods, but usually single-brooded in Britain.

SIMILAR SPECIES Eastern Wood White, of S France and SE Europe, has no white spot on antenna.

♀ ▲

Mountain Clouded Yellow

Colias phicomone

A fast-flying butterfly with conspicuous red fringes on all wings. The upperside, rarely seen at rest, is pale yellow in the male and almost white in the female, but both sexes are heavily dusted with black. The dark outer border contains several indistinct pale spots. Winter is passed as a caterpillar.

SIZE 24 mm.

HABITAT Flowery mountain slopes, usually above 1,800 m.

FOODPLANT Alpine milk-vetch and other legumes.

Range Alps, Pyrenees and mountains of N Spain.

FLIGHT June–September, in one or two broods.

SIMILAR SPECIES Pale Arctic Clouded Yellow is very similar, but is confined to far N. Moorland Clouded Yellow (opposite) has solid black border and almost unmarked underside.

caterpillar

A fast-flying, restless butterfly, rarely settling for long and hardly ever opening its wings at rest. The upperside of the male varies from lemon yellow to white, while that of the female is always white. Both sexes have a broad black border, without pale spots, and red fringes. The underside is largely unmarked. Winter is passed as a caterpillar or egg. One of the commonest butterflies of the Arctic.

SIZE 26 mm.

HABITAT Bogs and moorland.

FOODPLANT Bilberry and bog bilberry.

RANGE N & C Europe, from the Alps eastwards.

♀

▲

FLIGHT June–July.

SIMILAR SPECIES Mountain Clouded Yellow (opposite) has red-ringed spot under hindwing.

Pale Clouded Yellow

Colias hyale

A fast-flying butterfly, hardly ever opening its wings at rest. Look for the deep yellow underside of the hindwing and the paler forewing – lemon in the male and white in the female. The upperside is lemon or white with pale spots in the dark border. Winter is passed as a caterpillar. A rare summer visitor to Britain.

SIZE 24 mm.

HABITAT Flower-rich grassland, especially lucerne and clover fields.

FOODPLANT Lucerne, vetches and other legumes.

RANGE S & C Europe; migrates northwards in spring and summer.

FLIGHT May–September, in two broods.

SIMILAR SPECIES Berger's Clouded Yellow is almost identical, but its caterpillar is very different.

♀

A fast-flying butterfly, rarely opening its wings at rest. Both sexes are golden yellow on both surfaces. The upperside has a wide black border, that of the female enclosing pale spots. About 10% of females are pale, like the Pale Clouded Yellow (opposite), but with a broader black border. Winter is passed as a caterpillar.

SIZE 25 mm.

HABITAT Clover and lucerne fields; other flowery places.

FOODPLANT Clovers, vetches and other legumes.

RANGE Almost all Europe, but only as summer visitor north of Alps.

FLIGHT April–October, in several broods.

SIMILAR SPECIES Northern Clouded Yellow of the Arctic has dusky underside. Several others in SE Europe.

The male upperside is bright yellow, but the female is greenish white and like a Large White (p.35) in flight. The tip of the forewing is drawn out to a sharp point and all four wings have a red dot in the centre. The underside is pale green and leaflike in both sexes. Adult overwinters, hiding in ivy and other evergreens, and is among the first butterflies to reappear in spring.

SIZE 30 mm.

HABITAT Open woodland, hedgerows and gardens.

FOODPLANT Buckthorn and alder buckthorn.

RANGE Almost all Europe except Scotland.

♀

FLIGHT June–September; again in spring after hibernation.

SIMILAR SPECIES Female Cleopatra (opposite) has reddish streak on underside of forewing.

Male upperside is brilliant yellow with a large orange patch on the forewing. The female is very pale yellow. The underside is pale greenish yellow and very leaflike in both sexes, although the female has a faint orange streak running through the forewing. The adult overwinters in evergreen trees and shrubs.

SIZE 30 mm.

HABITAT Open woodland and scrub.

FOODPLANT Buckthorn.

RANGE S Europe.

FLIGHT May–August and again in early spring after hibernation. May fly almost all year in S Spain, where it is double-brooded.

SIMILAR SPECIES Brimstone female (opposite) is greener with no orange streak on underside.

Black-Veined White

Aporia crataegi

A somewhat 'lazy' flier, with thinly scaled wings, especially in the female which has browner veins than the male. Upper and lower surfaces are alike and the wings often become transparent with age. Large numbers often gather to drink from puddles. Winter is passed as a caterpillar, in a communal silk nest.

SIZE 32 mm.

HABITAT Open country, woodland edges, gardens and orchards, where it is often a nuisance.

FOODPLANT Hawthorn, blackthorn and various fruit trees.

RANGE Most of Europe, but extinct in Britain.

FLIGHT May–August.

SIMILAR SPECIES None.

caterpillar

The familiar 'cabbage white', whose gregarious yellow and black caterpillars cause so much damage to brassica crops. A fast-flying butterfly. The male (above) lacks the black spots present on the upperside of the female (see right). The undersides are alike in both sexes, with yellowish hindwings and white forewings with two black spots. Winter is passed as a pupa.

SIZE 32 mm.

HABITAT Flowery places, especially on cultivated land.

FOODPLANT Mainly cultivated brassicas.

RANGE All Europe; often migrates in swarms.

FLIGHT April–October, in 1–3 broods.

SIMILAR SPECIES Small White (p.36) is smaller, with less black on wing-tips. Moorland Clouded Yellow (p.29) has more extensive black border.

A troublesome cabbage pest whose green larvae are very difficult to see on cabbage leaves. The female, shown here, has two black spots on the upperside of the forewing; the male has just a single spot. The underside has two black spots in both sexes, and yellowish hindwings. Winter is passed as a pupa.

SIZE 25 mm.

HABITAT Flowery places of all kinds; especially common in gardens and other cultivated areas.

FOODPLANT Cultivated brassicas and many wild crucifers.

RANGE All Europe.

FLIGHT March–October, in 2–4 broods.

SIMILAR SPECIES Large White (p.35) is larger, with denser black markings. Southern Small White of S & C Europe has more extensive black wing-tips.

The veins of the underside are edged with green, but this is often indistinct in summer broods. Females have two black spots on the upperside of the forewing; males have just one. The underside of the forewing has two spots in both sexes. Winter is passed as a pupa. This species is not a cabbage pest.

SIZE 25 mm.

HABITAT Flowery meadows, gardens and hedgerows; mainly in damp habitats in N.

FOODPLANT Wild crucifers, including watercress.

RANGE All Europe.

FLIGHT March–November, in 1–3 broods.

SIMILAR SPECIES Peak White of Alps and Pyrenees has similar veins, but forewing has black spot at end of cell.

Pontia daplidice

The upperside, rarely seen at rest, has black tips to the forewing enclosing white spots. The underside of the forewing has two large black spots near the outer edge and another on the front edge. The female has deeper black markings than the male, shown here. Winter is passed as a caterpillar or chrysalis.

♀

SIZE 23 mm.

HABITAT Rough, flowery places.

FOODPLANT Various wild crucifers; also mignonette.

RANGE Most of Europe except far N. A great migrant, but rare vagrant to British Isles.

FLIGHT February–October, in 2–4 broods.

SIMILAR SPECIES Several in S Europe, but none with two black spots near outer edge of underside of forewing.

The male, shown here, is unmistakable. The female has no orange and has more-rounded wing-tips with a solid black patch on the upperside. The underside of the hindwing is mottled white and green in both sexes. Winter is passed as a chrysalis.

SIZE 23 mm.

HABITAT Hedgerows, gardens and damp meadows.

FOODPLANT Garlic mustard, cuckooflower and other wild crucifers; also mignonette, garden honesty and sweet rocket.

♀

RANGE Most of Europe except far N.

FLIGHT April–July.

SIMILAR SPECIES Female is like female Bath White (opposite), but latter has two black spots on underside of forewing.

Anthocharis belia

The male, seen here, is yellow on both surfaces, but the upperside has a larger orange patch on the forewing and lacks the black mottling on the underside of the hindwing. The female is largely white, with a dirty orange tip to the forewing and yellow and black mottling on the underside of the hindwing. Winter is passed as a pupa.

SIZE 20 mm.

HABITAT Flowery fields, usually in uplands.

FOODPLANT Buckler mustard and some other crucifers.

RANGE SW Europe.

FLIGHT March–June.

SIMILAR SPECIES Eastern Orange-tip of SE Europe has similar male, but female has black wing-tips.

caterpillar

A sturdy, fast-flying butterfly fond of feeding at ripe fruit. There are two 'tails' on each hindwing and an intricate pattern on the underside. The upperside is velvety brown with orange margins. Winter is passed as a caterpillar.

SIZE 40 mm.

HABITAT Scrubby places and orchards.

FOODPLANT Strawberry tree.

RANGE Mediterranean, rarely far from the coast.

FLIGHT May–October, in two broods.

SIMILAR SPECIES None, although Camberwell Beauty (p.52) is similar in flight.

caterpillar

Purple Emperor

Apatura iris

A strong flier living mainly in the treetops, although the male, seen here, often descends to drink from muddy puddles and carrion. The brilliant purple sheen, visible only from certain angles, is absent in the female. The underside is largely brown and grey in both sexes, with a white stripe through the centre. Winter is passed as a very small caterpillar.

SIZE 37 mm.

HABITAT Mature woodland, usually with oaks.

FOODPLANT Sallows.

RANGE S & C Europe, including S England; absent from most of Mediterranean.

FLIGHT June–August.

SIMILAR SPECIES Lesser Purple Emperor (opposite) has orange-ringed black spot on forewing.

A fast-flying butterfly, often gliding for long distances. The upperside is brown, crossed by a broken white or yellowish stripe and with a deep bluish-purple iridescence in the male. Near the outer edge is a black spot ringed with orange. The underside is largely orange-brown with black and white spots. Winter is passed as a small caterpillar.

SIZE 35 mm.

HABITAT Light woodland.

FOODPLANT Willows and poplars, especially aspen.

RANGE S & C Europe, except British Isles and most of Iberia.

FLIGHT May–September, in one or two broods.

SIMILAR SPECIES Purple Emperor (opposite) has no orange ring around black spot on forewing.

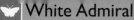

The velvety black upperside, crossed by a broken white stripe, becomes browner with age. The underside is orange-brown, with a white stripe and two rows of black dots near the edge of the hindwing. Flight is quite slow, with elegant glides, and the butterfly likes to bask in dappled shade. It is very fond of bramble blossom. Winter is passed as a small caterpillar.

SIZE 30 mm.

HABITAT Woodland rides and clearings, often in quite shady areas.

FOODPLANT Honeysuckle.

RANGE S & C Europe, including S Britain.

FLIGHT June–August.

SIMILAR SPECIES Southern White Admiral (opposite) has white spot near middle of forewing.

caterpillar

A rather slow butterfly with a graceful gliding flight. The velvety black upperside is crossed by a broken white stripe and has a white spot near the middle of the forewing. The underside is rich brown, with a large silvery patch and a single row of black dots on the hindwing. Winter is passed as a small caterpillar.

SIZE 27 mm.

HABITAT Light woodland and surrounding scrub.

FOODPLANT Honeysuckle.

RANGE S & C Europe except British Isles.

FLIGHT May–September, in 1–3 broods.

SIMILAR SPECIES White Admiral (opposite) has no white spot in middle of forewing.

Limenitis populi

A powerful flier, seen mainly in the treetops although it may come down to feed at dung or carrion. The upperside is dull brown with white spots and stripes and a conspicuous row of orange crescents on the hindwing. The underside is largely orange, with white spots and a greyish-blue margin. Winter is passed as a small caterpillar.

SIZE 40 mm.

HABITAT Light woodland, especially near water.

FOODPLANT Aspen and other poplars.

RANGE Much of Europe, except British Isles, Iberia and Mediterranean.

FLIGHT June–August.

SIMILAR SPECIES
Female Purple Emperor (p.42) has no orange crescents on hindwing.

caterpillar

A dainty, gliding butterfly brownish black above and chocolate brown below. Both surfaces are crossed by a broad white band, although this is broken on the forewing. The latter also has a white spot near the centre. Winter is passed as a caterpillar.

SIZE 26 mm.

HABITAT Open woodland.

FOODPLANT Goatsbeard spiraea and meadowsweet.

RANGE E Europe, as far west as Switzerland.

FLIGHT June–July.

SIMILAR SPECIES
Southern White Admiral (p.45) has more pointed hindwing, with large silvery patch on underside. Common Glider has two white bands on hindwing.

caterpillar

The velvety black upperside with its red and white markings is unmistakable. The underside is similar, but paler and more mottled. The butterfly is very fond of ripe fruit. The adult overwinters, but few survive the cold in Britain and Northern Europe; these areas are repopulated each year by migration from the south.

SIZE 30 mm.

HABITAT All flower-rich places, including gardens.

FOODPLANT Stinging nettle.

RANGE All Europe; regularly reaches Iceland.

FLIGHT May–October, usually in two broods, and again in early spring after hibernation.

SIMILAR SPECIES
None.

caterpillar

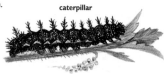

This fast-flying butterfly is a great migrant. It cannot survive the European winter, but arrives from North Africa each spring and spreads to all parts of Europe, including Iceland. It basks with its wings wide open, displaying its orange and black upperside. Look for the round black dots near the edge of the hindwing. The underside is similar, but paler and more mottled.

SIZE 30 mm.

HABITAT Any flowery habitat, including parks and gardens.

FOODPLANT Thistles; sometimes nettles and mallows.

RANGE All Europe.

FLIGHT April–October, in two broods.

SIMILAR SPECIES The tortoiseshells (pp.50 & 51) are darker, with no black dots on hindwing.

caterpillar

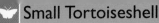

One of Europe's commonest garden species, and one of the earliest to appear in the spring. Look for the blue-spotted margins and the large dark patch at the base of the hindwing. The underside is black and brown with a faint blue border, affording ideal camouflage in its overwintering sites which include sheds, attics and hollow trees.

SIZE 25 mm.

HABITAT Gardens and other flowery places.

FOODPLANT Stinging nettle.

RANGE All Europe; occasionally reaches Iceland.

FLIGHT May–October, in 1–3 broods, and again in early spring after hibernation.

SIMILAR SPECIES Large Tortoiseshell (opposite) is duller, with no white near tip of forewing.

caterpillars

The upperside is dull orange with black spots but not much black on the hindwing. Look for the blue-spotted border. The underside is mottled brown, with a blue border to both wings. The adult overwinters in attics, outbuildings, hollow trees and log-piles. It is fond of drinking sap from injured trees.

SIZE 30 mm.

HABITAT Light woodland, orchards and gardens.

FOODPLANT Elms, sallows and other deciduous trees.

RANGE S & C Europe; probably extinct in British Isles other than as rare visitor.

FLIGHT June–August; again in spring after hibernation.

SIMILAR SPECIES Small Tortoiseshell (opposite) is brighter, with white spots on forewing. Painted Lady (p.49) has a lot of white on forewing.

caterpillar

Nymphalis antiopa

The underside is sooty brown with a dirty white border. The adult overwinters, tucked up in a hollow tree or other dark place where its sombre underside camouflages it well. Although a great migrant, it is a scarce vagrant to the British Isles, usually arriving in autumn.

SIZE 33 mm.

HABITAT Lightly wooded areas.

FOODPLANT Sallows and willows.

RANGE Almost all Europe, but a rare vagrant to British Isles.

caterpillar

FLIGHT June–September, and again in spring after hibernation.

SIMILAR SPECIES None, although Two-tailed Pasha (p.41) looks similar in flight.

A fast-flying butterfly, very fond of garden buddleias in summer. It is a strong migrant. It perches with wings open or closed and often flashes its eye-spots to scare predators. Winter is passed as an adult, often in dark buildings, where the almost black underside provides perfect camouflage.

caterpillar

SIZE 30 mm.

HABITAT Gardens, wasteland and other flower-rich places.

FOODPLANT Stinging nettle.

RANGE All Europe except far N; occasionally reaches Iceland.

FLIGHT June–September, and again in spring after hibernation.

SIMILAR SPECIES None.

The 'ragged' wings are mottled brown below and resemble dead leaves. The underside of the hindwing bears the white, comma-shaped mark that gives the species its name. Late summer insects are darker than the midsummer one shown here. The adult overwinters in hedges and undergrowth, its leaf-like wings providing perfect camouflage.

1st brood

SIZE 24 mm.

HABITAT Light woodland, hedgerows and gardens.

FOODPLANT Hops, stinging nettle and elm.

RANGE Most of Europe, except Ireland and N Britain.

FLIGHT June–September, in two broods, and again in spring after hibernation.

SIMILAR SPECIES Southern Comma of S Europe has smaller spots and white V instead of 'comma'.

Named for the irregular lines, like map boundaries, on its underside.
The spring brood has an orange upperside with black spots, but
summer insects are dark brown with a broken cream or white
stripe. Flight consists mainly of short glides close to the ground or
vegetation. Winter is passed as a pupa.

SIZE 18 mm.

HABITAT Open woodland and other rough ground.

FOODPLANT Stinging nettle.

RANGE C Europe and N Iberia; absent from British Isles.

FLIGHT April–September, in two or three broods.

SIMILAR SPECIES None, but spring insects are somewhat like
small fritillaries and summer
ones are like White
Admiral (p.44) in flight.

caterpillar

Argynnis paphia

The female upperside is duller than the male with larger spots, but without the black streaks on the forewing – these carry the scent scales. The underside of the hindwing is pale green with a purplish margin and the silver streaks responsible for the insect's name. Winter is passed as a tiny caterpillar.

SIZE 35 mm.

HABITAT Deciduous woodland and surrounding scrub.

FOODPLANT Violets.

RANGE Most of Europe, except Scotland and far N.

FLIGHT June–September.

SIMILAR SPECIES Cardinal (opposite) has green-tinted upperside. Dark Green Fritillary (p.58) has silver spots on underside.

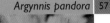

The upperside is orange with black spots and a distinct green tinge in both sexes. The underside of the forewing is largely peach-coloured with black spots, while the underside of the hindwing is pale green with faint silvery stripes, best seen in the female. A fast-flying butterfly with a great liking for thistles. Winter is passed as a caterpillar.

SIZE 32–40 mm.

HABITAT Flower-rich grassland and woodland clearings.

FOODPLANT Violets.

RANGE S Europe, as far N as Slovakia in E; also Atlantic coast of France.

FLIGHT May–July.

SIMILAR SPECIES Silver-washed Fritillary (opposite) is more orange above and lacks peach colour below..

A fast-flying butterfly with a variable pattern of black spots on its rich brown upperside. On the underside, the hindwing and the tip of the forewing are greenish with silver spots. The butterfly is particularly fond of thistles and knapweeds. Winter is passed as a small caterpillar.

SIZE 30 mm.

HABITAT Rough grassland, heaths, coastal dunes and open woodlands.

FOODPLANT Violets.

RANGE All Europe.

FLIGHT June–August.

SIMILAR SPECIES
High Brown Fritillary has row of silver-centred red spots on underside of hindwing, and less green.

caterpillar

The upperside is rich orange with black spots and the outer edge of the forewing is slightly concave. The underside of the hindwing has very large silver spots. The species can pass the winter in any stage from egg to adult, but usually as a caterpillar or chrysalis.

SIZE 20–25 mm.

HABITAT Flowery fields and hillsides.

FOODPLANT Violets.

RANGE All Europe, but rare summer visitor to British Isles.

FLIGHT February–November, in 1–3 broods.

SIMILAR SPECIES No other fritillary has such large and shiny silver spots on underside.

caterpillar

The relatively short, rounded wings are bright orange above, with far from uniform black spots. Look for two particularly large spots just beyond the middle of the forewing. The underside of the hindwing is strongly tinged with purple in the outer half, especially so in the female, in which the wing is virtually half yellow and half purple. Winter is passed as a caterpillar.

SIZE 21–25 mm.

HABITAT Rough, flower-rich grassland.

FOODPLANT Violets and bramble.

RANGE S & C Europe except British Isles.

FLIGHT June–August.

SIMILAR SPECIES
Lesser Marbled Fritillary is
slightly smaller, with less
purple on underside.

caterpillar

The upperside is of the usual orange fritillary colour, but the black spots in the basal half of the forewing are distinctly V-shaped. The underside of the forewing has much darker spotting than most similar fritillaries, and the underside of the hindwing is largely rust-red with silvery spots around the edge. Winter is passed as a caterpillar.

caterpillar

SIZE 18 mm.

HABITAT Bogs and wet heaths.

FOODPLANT Cranberry.

RANGE N & C Europe except British Isles.

FLIGHT June–July.

SIMILAR SPECIES Shepherd's Fritillary of mountainous areas has yellow patch near outer edge of underside of hindwing.

Named for the silvery spots, edged with brown, on the outer margin of the underside of the hindwing. There is also a large silvery spot in the centre of this wing. The upperside has the orange and black pattern typical of the fritillaries. Winter is passed as a caterpillar.

SIZE 22 mm.

HABITAT Open woodland, especially coppiced areas, and the surrounding fields and scrub.

FOODPLANT Violets.

RANGE Almost all Europe.

FLIGHT April–August, in one or two broods.

SIMILAR SPECIES Small Pearl-bordered Fritillary (opposite) normally has band of silvery spots across underside of hindwing, and marginal spots are edged with black.

caterpillar

The central band running across the underside of the hindwing is usually all silver, although sometimes yellow. Look for the black edging to the silvery marginal spots. The upperside is virtually identical to that of the Pearl-bordered Fritillary. Winter is passed as a caterpillar.

SIZE 20 mm.

HABITAT Open woodland, moors and damp grassland.

FOODPLANT Violets.

RANGE Almost all Europe, except Ireland and Mediterranean coast.

FLIGHT May–August, in one or two broods.

SIMILAR SPECIES Pearl-bordered Fritillary (opposite) has single silver spot in centre of underside of hindwing.

caterpillar

One of the smallest fritillaries, also called Violet Fritillary. The hindwing, sharply angled at its tip, is purplish brown below with silvery marginal spots and large silver spots in a central band. The upperside displays the normal fritillary pattern. It usually flits and glides close to the ground. Winter is passed as a caterpillar.

SIZE 18 mm.

HABITAT Rough grassland, heaths and light woodland.

FOODPLANT Violets and low-growing brambles.

RANGE S & C Europe except British Isles.

FLIGHT April–October, with two or three broods.

SIMILAR SPECIES None outside the Arctic.

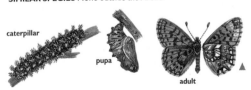

caterpillar

pupa

adult

Easily recognized by the bright yellow band on the underside of the hindwing, which also has metallic grey marginal spots. The upperside is heavily spotted with brown in the Alpine race and, as shown here, the hindwing may be almost entirely brown. The Scandinavian race is much paler and the yellow band on the underside is much less obvious. The species spends two winters as a caterpillar.

SIZE 22 mm.

HABITAT Montane woodland and northern birch scrub.

caterpillar

FOODPLANT Violets, including yellow wood violet.

RANGE Alps and Arctic Scandinavia.

FLIGHT June–July.

SIMILAR SPECIES Yellow band on hindwing should always distinguish this species.

Glanville Fritillary

Melitaea cinxia

The orange upperside is crossed by conspicuous brown lines, and there is an arc of black spots near the outer edge of the hindwing. The underside of the hindwing is creamy white with two irregular orange bands, the outer one containing darker spots. Winter is passed as a caterpillar in a communal silk nest.

SIZE 15–20 mm.

HABITAT Flower-rich grasslands, including roadsides.

FOODPLANT Plantains; occasionally hawkweeds and knapweeds.

RANGE Most of Europe except far N; only on Isle of Wight in British Isles.

FLIGHT May–September, in one or two broods.

SIMILAR SPECIES Heath Fritillary (p.69) has similar upperside, but no spots near edge of hindwing.

caterpillar

Variable in size and pattern, but the upperside shows a conspicuous mixture of light and dark orange. There are no black spots near the edge of the hindwing. Look for the much-enlarged marginal crescent three spaces up from the rear of the forewing. The underside of the hindwing is largely cream, with an orange band containing red spots near the outer edge. Winter is passed as a caterpillar in a communal silk nest.

SIZE 24 mm.

HABITAT Flower-rich grassland.

caterpillar

FOODPLANT Knapweed
and sometimes plantains.

RANGE S & C Europe
except British Isles.

FLIGHT April–September, in 1–3 broods.

SIMILAR SPECIES Freyer's Fritillary of SE Europe has black spots near outer edge of hindwing.

A very variable butterfly. In its typical form, the male upperside is bright orange with scattered small black spots, while the female is paler, with heavier spots and usually thickly dusted with grey scales. The underside is more constant, with two clear orange bands on the cream hindwing. Winter is passed as a caterpillar in a communal silk shelter.

SIZE 17–23 mm.

HABITAT Flowery grassland.

FOODPLANT Plantains, speedwells and toadflaxes.

RANGE S & C Europe except British Isles.

caterpillar

FLIGHT April–September, in two or three broods.

SIMILAR SPECIES Lesser Spotted Fritillary has clearly triangular (not rounded) black spots near edge of underside of hindwing.

In the northern half of Europe the upperside has heavy brown lines, as shown here. In the south the ground colour is often much paler and the markings are lighter. The underside is largely orange, with white or cream bands on the hindwing: look for the heavy black edges to the marginal lunules or crescents near the rear of the forewing. Winter is passed as a caterpillar in a communal web.

SIZE 20 mm.

HABITAT Open woodland, scrub and flowery grassland.

FOODPLANT Common cow-wheat and ribwort plantain.

RANGE Most of Europe, but very rare in S England.

FLIGHT May–October, in 1–3 broods.

caterpillar

SIMILAR SPECIES Hard to split from Provençal Fritillary (p.70) and Knapweed Fritillary (p.67) in S.

The upperside often has two shades of orange, with fairly light and often broken brown lines. The underside of the forewing has no heavy black edging to the lunules in the rear half, but there is often a dumb-bell-shaped mark near the rear edge. The underside of the hindwing is cream with orange bands, the outer one often containing red spots. Winter is passed as a half-grown caterpillar.

SIZE 22 mm.

HABITAT Flowery places, usually in uplands.

FOODPLANT Toadflaxes.

RANGE SW Europe, from Portugal to Alps.

FLIGHT May–September, in one or two broods.

SIMILAR SPECIES Heath Fritillary (p.69) has heavy black edges to marginal lunules under forewing.

The male, shown here, is Europe's only fritillary with white patches on the upperside. At high altitudes it is almost entirely black and white. The white areas are replaced by orange in the female, but the darker orange or red band remains conspicuous in the outer part of each wing and may contain black spots on the hindwing. The underside is largely yellow with brick-red bands. Winter is passed as a caterpillar in a communal web.

SIZE 20 mm.

HABITAT Montane scrub and grassland.

FOODPLANT Plantains, lady's mantles and *Viola* species.

RANGE Alps and Bulgarian mountains.

FLIGHT May–August.

SIMILAR SPECIES Female like Asian Fritillary of Alps, but this lacks pale lunules on upperside of hindwing.

pupa

caterpillar

Upper and lower surfaces are similarly marked with orange and yellow bands. The pattern may vary, but there is always a distinct row of black spots on both sides of the hindwing. There are no black spots on the underside of the forewing. Winter is passed as a caterpillar in a communal silk shelter.

SIZE 20 mm.

HABITAT Moors and rough grassland, both wet and dry.

FOODPLANT Scabious, especially devil's-bit scabious, plantains and yellow gentian.

RANGE All Europe, except E Britain and far N.

FLIGHT April–July.

SIMILAR SPECIES Spanish Fritillary is much brighter and has bold black marks under the forewing.

Look for the eye-spots on the underside to distinguish this slow-flying butterfly from members of the White Family. The upperside is chequered black and white and the black markings are often very extensive on butterflies in the mountains. The bands on the underside of the hindwing are grey or yellowish. The female scatters her eggs in flight. Winter is passed as a small caterpillar.

SIZE 22–28 mm.

HABITAT Rough, flowery grassland.

FOODPLANT Red fescue and other grasses.

RANGE S & C Europe, except Scotland and Ireland.

FLIGHT June–August.

SIMILAR SPECIES Several in S Europe: Western Marbled White (p.74) has brown veins on underside.

caterpillar

The veins on the underside of the hindwing are brown and the eye-spots, usually clearly visible on both surfaces of the hindwing, have faint blue centres. A thick black bar crosses the cell on the forewing. Eggs are scattered freely over the grass and winter is passed as a caterpillar.

SIZE 27 mm.

HABITAT Rough grassland, usually in uplands.

FOODPLANT Assorted grasses.

RANGE SW Europe, from Portugal to W Italy.

FLIGHT May–July.

SIMILAR SPECIES Marbled White (p.73) is normally more heavily marked. Several others in S Europe all have black veins.

The upperside, rarely seen, is dull brown with a white or yellowish outer band, which is dusted with brown in the male. The inner edge of the band on the hindwing is almost straight or gently curved, and lightly toothed. The underside of the forewing has a pale band with one or two eye-spots. The butterfly generally rests on tree-trunks. Winter is passed as a caterpillar.

SIZE 33–38 mm (very rarely under 33 mm).

HABITAT Light woodland and scrub.

FOODPLANT Assorted grasses.

RANGE S & C Europe, except British Isles and most of Iberia.

FLIGHT June–September.

SIMILAR SPECIES Rock Grayling (p.76) is almost identical, but forewing usually under 33 mm long.

Hipparchia alcyone

The upperside is greyish brown with a pale yellowish band, heavily dusted with brown in the male. The inner edge of the band on the hindwing is quite strongly curved. The underside of the forewing resembles the upperside. The butterfly normally rests on rocks. Winter is passed as a caterpillar.

SIZE 30–33 mm (very rarely over 33 mm).

HABITAT Rocky slopes, usually in uplands.

FOODPLANT Assorted grasses.

RANGE S & C Europe and S Norway.

FLIGHT June–August.

SIMILAR SPECIES Woodland Grayling (p.75) is almost identical, but forewing is rarely under 33 mm long. The two caterpillars are quite distinct.

Agile and usually very difficult to spot when settled on the ground
with its wings closed and keeled over towards the sun to eliminate
shadows. The rarely-seen upperside is dull brown with a paler band
in the outer region. The band on the hindwing carries a number of
fairly distinct orange squares or triangles. There are two eye-spots
on each surface of the forewing. Winter is passed as a caterpillar.

SIZE 20–30 mm.

HABITAT Heathland, coastal dunes
and rough grassland.

FOODPLANT Assorted fine-leaved grasses.

RANGE Most of Europe except far N.

FLIGHT June–September.

SIMILAR SPECIES Several in S Europe, but all lack orange patches
in pale band of hindwing.

Hipparchia statilinus

The upperside is dull brown, often almost black in the male. Two large eye-spots are visible on both sides of the forewing, separated by two small white spots. The eye-spots of the female usually have white pupils, but those of the male are usually blind. The underside of the hindwing is often unmarked. Winter is passed as a caterpillar.

SIZE 23 mm.

HABITAT Heathland, light woodland and scrub.

FOODPLANT Assorted grasses.

RANGE S & C Europe except British Isles.

FLIGHT June–September.

SIMILAR SPECIES Striped Grayling of SW Europe has black and white stripes under hindwing.

The upperside is dark brown with a cream band on each wing. The forewing has a white or cream front edge and two eye-spots in the band. The underside pattern varies a lot and the underside of the hindwing may be unmarked in the female. Look for the dark, often triangular spot in the cell of the underside of the forewing. Winter is passed as a caterpillar.

SIZE 20–32 mm (very variable).

HABITAT Dry, stony grassland.

FOODPLANT Blue moor grass and other fine grasses.

RANGE S & C Europe except British Isles.

FLIGHT June August.

SIMILAR SPECIES Southern Hermit from C Spain has prominent pale patch near middle of upperside of forewing.

Both surfaces of the male, shown here, are sooty brown, but the female is much paler and her forewing is largely orange below. Both sexes have two eye-spots on the forewing, those of the female on an orange patch. The eye-spots appear on both surfaces, normally separated by two white dots. Winter is passed as a caterpillar.

SIZE 25–30 mm.

HABITAT Rough grassland and stony slopes.

FOODPLANT Fescues and other fine grasses.

pupa

RANGE S & C Europe from Pyrenees eastwards, except British Isles.

FLIGHT June–August.

SIMILAR SPECIES Black Satyr of SW Europe generally has two white bands under hindwing and male has only one eye-spot on forewing.

caterpillar

Each forewing carries two blue-centred eye-spots. The male, shown here, is chocolate brown on both surfaces, with a darker line near the edge of the underside of the hindwing. The female is lighter brown, often with two white stripes on the underside of the hindwing. Winter is passed as a caterpillar.

SIZE 28–35 mm.

HABITAT Light woodland and rough grassland.

FOODPLANT Various grasses, including purple moor grass.

RANGE S & C Europe, except British Isles and most of Spain and Italy.

FLIGHT July–September.

SIMILAR SPECIES Several are superficially similar, but none has blue-centred eye-spots.

The upperside is dark brown in both sexes, with a clear white or cream band on each wing. The band is broken into spots on the forewing and contains a single eye-spot, which is usually blind. The underside is similar, with additional white spots near the middle of the forewing and a short stripe near the base of the hindwing. The butterfly often basks on rocks and roads. Winter is passed as a caterpillar.

SIZE 33–40 mm.

HABITAT Light woodland and rough grassland.

FOODPLANT Various grasses.

RANGE S & C Europe except British Isles.

caterpillar

FLIGHT June–August.

SIMILAR SPECIES Hermit (p.79) is smaller with two eye-spots on forewing.

The upperside is velvety brown and all wings usually have shiny orange patches containing a variable number of small black dots. The underside ranges from pale to deep brown, usually with a fair amount of orange on the forewing. The butterfly basks with its wings wide open in the sunshine, but hides away at other times. Winter is passed as a small caterpillar.

SIZE 19 mm.

HABITAT Moors and upland grassland.

caterpillar

FOODPLANT Mat grass and some other grasses.

RANGE Mountains of S & C Europe and N Britain.

FLIGHT June–August.

SIMILAR SPECIES Scotch Argus (p.84) and many others in mountains on the Continent.

The upperside is velvety brown with a rust-coloured band containing several white-centred eye-spots on each wing. The underside of the forewing resembles the upperside, but the underside of the hindwing has one or two silvery-grey or yellowish bands, the outermost containing four small eye-spots. The butterfly basks in the sunshine, but hides in the vegetation in dull weather. Winter is passed as a caterpillar.

SIZE 25 mm.

HABITAT Moors and open woodland, mainly in uplands.

FOODPLANT Purple moor grass and other grasses.

RANGE N Britain and C Europe.

FLIGHT July–September.

SIMILAR SPECIES Mountain Ringlet (p.83) lacks white-centred eye-spots. Several more on the Continent.

Both surfaces are chocolate brown with distinctly almond-shaped orange spots, most containing tiny black eye-spots with minute white pupils. The orange spots vary a lot in size, but are usually clear enough for accurate identification. Winter is passed as a caterpillar.

SIZE 22 mm.

HABITAT Montane grassland, usually above 900 m.

FOODPLANT Fescues and other fine-leaved grasses.

RANGE Alps and some other mountains of S Europe, but apparently not in Pyrenees.

caterpillar

FLIGHT June–August.

SIMILAR SPECIES
Several are superficially similar, but none has almond-shaped spots.

♂

Dewy Ringlet

Erebia pandrose

The upperside is chocolate brown and the forewing bears a more or less rectangular brick-red patch, bounded internally by a dark line and containing four blind black eye-spots. The hindwing may have four red-ringed spots. Look for the two dark, wavy lines on the underside of the grey hindwing, sometimes enclosing a dark brown band. Winter is passed as a caterpillar.

SIZE 20–25 mm.

HABITAT Montane slopes and northern tundra.

FOODPLANT Fescues and other fine-leaved grasses.

RANGE Mountains and tundra of Scandinavia and mountains of S & C Europe.

FLIGHT June–August.

SIMILAR SPECIES False Dewy Ringlet lacks dark lines on underside of hindwing.

caterpillar

One of Europe's commonest butterflies. The male upperside, shown here, is dark brown with a dark scent patch in the middle of the forewing and no more than a faint orange smudge. The underside of the forewing is largely orange in both sexes, and the hindwing is mottled brown. Winter is passed as a small caterpillar.

SIZE 25 mm.

HABITAT Grassland of all kinds.

FOODPLANT Various grasses, especially *Poa* species.

RANGE All Europe except far N.

FLIGHT May–September.

SIMILAR SPECIES Dusky Meadow Brown and Oriental Meadow Brown have blind eye-spots on upperside: usually one in male and two in female.

Ringlet

Aphantopus hyperantus

The upside is deep blackish brown and velvety when fresh. The eye-spots are often obscure and may be absent in males. The conspicuous cream-ringed eye-spots on the otherwise unmarked brown underside readily identify this common butterfly, which is very fond of bramble blossom. Winter is passed as a small caterpillar.

SIZE 19 mm.

HABITAT Woodland rides and shady hedgerows, especially in damp places with lush grass.

FOODPLANT Various grasses.

RANGE Most of Europe, except far N and S.

FLIGHT June–August.

SIMILAR SPECIES False Ringlet (p.93) has paler underside, with silvery line beyond eye-spots.

caterpillar

Also called Hedge Brown. The sexes are similar, but the female lacks the dark scent patch seen here on the male forewing. The eye-spot on the forewing usually has two white pupils. The underside of the forewing is largely orange and the hindwing is yellow and brown with brown-ringed white spots. The butterfly is very fond of bramble blossom. Winter is passed as a caterpillar.

SIZE 19 mm.

HABITAT Hedges, woodland edges and scrubby grassland.

FOODPLANT Various grasses.

RANGE S & C Europe except Scotland.

FLIGHT July–September.

SIMILAR SPECIES Spanish Gatekeeper has yellow stripe under hindwing. Southern Gatekeeper is grey and brown on underside of hindwing.

♀

The upperside, never exposed at rest, is orange-brown with a variable amount of grey dusting. The underside of the forewing is largely orange, but the hindwing is almost entirely grey. Eye-spots may be well developed on both surfaces, but are small or absent in many northern areas, including N Scotland. Passes winter as a small caterpillar.

♀

SIZE 15–20 mm.

HABITAT Bogs, moors and damp grassland.

FOODPLANT Various sedges and grasses.

RANGE C & N Europe from Alps northwards, except S England.

FLIGHT June–August.

SIMILAR SPECIES Balkan Heath has less white on underside.

One of Europe's commonest butterflies. The upperside, never exposed at rest, is orange-brown with grey margins and a small eye-spot at the tip. The underside of the forewing is also largely orange, with a prominent eye-spot. The pale band and eye-spots under the hindwing are variably developed. Winter is passed as a small caterpillar.

SIZE 14–20 mm.

HABITAT Rough grassland and heathland.

FOODPLANT Assorted fine-leaved grasses.

RANGE All Europe except far N.

FLIGHT April–October, in 1–3 broods.

SIMILAR SPECIES Large Heath (opposite) is larger, with no obvious grey margins. Pearly Heath (p.92) and others have conspicuous eye-spots under hindwing.

caterpillar

Pearly Heath

Coenonympha arcania

The upperside of the forewing is orange with a wide brown border. The upperside of the hindwing is dull brown. The upperside is never exposed at rest. Look for the silvery marginal band on the underside of the hindwing, and the prominent anterior eye-spot, which is not enclosed in the pale band. Winter is passed as a caterpillar.

SIZE 19 mm.

HABITAT Grassland and open woods, mainly in uplands.

FOODPLANT Various grasses, especially melicks.

RANGE S & C Europe & S Sweden, but not British Isles.

FLIGHT June–September, in one or two broods.

SIMILAR SPECIES Dusky Heath is darker, with arc of eye-spots curving inwards. Alpine Heath has anterior eye-spot enclosed in pale band.

caterpillar

The upperside, never seen at rest, is dark brown, with faint eye-spots in the female. The underside has conspicuous eye-spots on both wings in the female, but the male, seen here, usually has prominent eye-spots only on the hindwing. Look for the silvery sub-marginal line beyond the eye-spots. Although widely distributed, the species is seriously threatened by drainage of its habitats. Winter is passed as a caterpillar.

SIZE 20 mm.

HABITAT Damp grassland, bogs & damp, open woodland.

FOODPLANT Grasses, sedges and yellow iris.

RANGE S & C Europe; local and rapidly declining.

FLIGHT June–July.

SIMILAR SPECIES Ringlet (p.88) is darker below, lacking silvery sub-marginal line.

caterpillar

Speckled Wood

Pararge aegeria

In northern and eastern Europe, including the British Isles, the spots are cream, as shown here, but in south-west Europe they are orange. The butterfly often basks on sun-dappled leaves with its wings open, but with its wings closed its mottled brown underside looks very like a dead leaf. The butterfly rarely visits flowers and feeds mainly on honeydew. Winter is passed as a caterpillar or chrysalis.

SIZE 21 mm.

HABITAT Shady woodland rides and clearings.

FOODPLANT Various grasses.

RANGE Most of Europe except far N.

FLIGHT March–October, in 1–3 broods.

SIMILAR SPECIES In SW Europe resembles Wall Brown (opposite), which is mainly grey under hindwings.

southern race

The male, shown here, resembles the female but has a thick brown scent patch in the centre of the forewing. The underside of the forewing resembles the upperside, but the underside of the hindwing is mottled grey with prominent eye-spots. The butterfly basks on rocks or on the ground with its wings wide open. Winter is passed as a caterpillar.

SIZE 23 mm.

HABITAT Rough grassy places with areas of bare ground.

FOODPLANT
Assorted coarse grasses.

RANGE S & C Europe and
S Scandinavia.

♀

FLIGHT March–October, in two or three broods.

SIMILAR SPECIES Speckled Wood (opposite) is similar in SW Europe, but underside of hindwing is browner. Large Wall Brown (p.96) is much darker.

The upperside is generally brown, with an orange patch containing one or more eye-spots on each wing, but in parts of south-west Europe the forewing is largely orange. The male has a dark scent patch on the forewing. The underside of the forewing resembles the upperside, but the underside of the hindwing is mottled grey. Winter is passed as a caterpillar.

SIZE 22–28 mm.

HABITAT Rough grassland, open woods; mainly upland.

FOODPLANT Assorted grasses.

RANGE Most of Europe, except far N and British Isles.

FLIGHT May–September, in one or two broods.

SIMILAR SPECIES Wall Brown (p.95) is usually brighter, with more cross lines in forewing cell.

The underside of this butterfly has pale stripes and prominent eye-spots and is instantly recognizable. Look also for the double yellow lines around the edges. The upperside is dull brown with large, yellow-ringed, blind eye-spots on each wing, as shown here. Winter is passed as a caterpillar.

SIZE 26 mm.

HABITAT Open woodland.

FOODPLANT Assorted grasses.

RANGE C Europe, including the Baltic region, but not British Isles.

FLIGHT June–August.

SIMILAR SPECIES
None.

caterpillar

Kirinia roxelana

The upperside, rarely seen at rest, is yellowish brown with a large orange-red patch on the forewing. The underside of the forewing is also largely orange, but the female has pale patches near the tip on both surfaces. The butterfly rarely flies far, even when disturbed. It feeds eagerly at the flowers of Christ's thorn bushes. Little is known of its life history.

SIZE 30 mm.

HABITAT Dry, scrubby grassland and old orchards.

FOODPLANT Grasses. Eggs often laid on tree-trunks.

RANGE SE Europe.

FLIGHT May–July.

SIMILAR SPECIES Large Wall Brown (p.96) has prominent eye-spots on forewing and hindwing is much greyer below.

The upperside is rich brown with orange patches. The underside of the forewing resembles the upperside, but the hindwing is greyish brown. Look for the long palps and the single tooth on the forewing. The butterfly resembles a dead leaf when resting with its wings closed. Winter is passed as an adult.

SIZE 20 mm.

HABITAT Lightly wooded areas, including orchards.

FOODPLANT Nettle tree.

RANGE S Europe, as far north as Slovakia in E.

FLIGHT May–August; again in spring after hibernation.

SIMILAR SPECIES Comma (p.54) is similar at rest, but has prominent tooth on hindwing and no orange on underside of forewing.

caterpillar

Hamearis lucina

Often called the Duke of Burgundy Fritillary because of its brown and orange coloration, this butterfly is easily distinguished from the smaller fritillaries by the two conspicuous white bands on the underside of the hindwing. Look also for the black spots all around the edges of the upperside. Flight is swift and usually close to the ground. Winter is passed as a pupa.

SIZE 14 mm.

HABITAT Scrubby hillsides and open woodland.

FOODPLANT Cowslip and primrose.

RANGE S & C Europe, including S Sweden, but not Scotland, Wales or Ireland.

FLIGHT May–September, in one or two broods.

SIMILAR SPECIES Chequered Skipper (p.13) is similar in flight, but has no black marginal spots.

The upperside, never seen at rest, is chocolate brown with no markings apart from a small oval scent patch at the front of the male forewing. There is often a line of white dots on the underside. The insect is well camouflaged at rest and the alternate flashing of brown and green makes it very hard to follow in flight. Winter is passed as a pupa.

SIZE 15 mm.

HABITAT Heaths, moors and scrubby grassland.

caterpillar

FOODPLANT Gorse, broom, heather, bilberry, rock-rose and many other shrubs and herbs.

RANGE All Europe.

FLIGHT March- July.

SIMILAR SPECIES Chapman's Green Hairstreak of SW Europe has complete white line on its underside.

Brown Hairstreak

Thecla betulae

The bright orange underside, striped with white and edged with red, makes this butterfly unmistakable. The upperside, rarely seen at rest, is dull brown in both sexes, but the female has a large red patch on the forewing shown here. Both sexes have small red patches on the 'tail' region of the hindwing. Winter is passed as an egg.

SIZE 18 mm.

HABITAT Woodland and scrub, usually flying fairly high up in trees.

FOODPLANT Blackthorn and other *Prunus* species.

RANGE Most of Europe, except Scotland and far N.

FLIGHT July–October.

SIMILAR SPECIES None.

The male upperside, seen here, is velvety black, shot all over with a rich purple sheen when seen from certain angles. The female is sooty black, the purple restricted to the base of the forewing. The underside is grey in both sexes, with a prominent white line on both wings and orange spots near the 'tail' on the hindwing. Winter is passed as an egg.

SIZE 14 mm.

HABITAT Mature woodland, usually keeping to treetops and feeding on honeydew.

FOODPLANT Oak.

RANGE Most of Europe except far N.

FLIGHT June–September.

SIMILAR SPECIES Spanish Purple Hairstreak has browner underside with orange border.

♀

Black Hairstreak

Strymonidia pruni

The upperside, never seen at rest, is dull brown with an orange border, although the latter may be absent from the forewing. The female sometimes has an extensive orange flush on the forewing. Look for the black spots bordering both sides of the orange band on the underside of the hindwing. Winter is passed as an egg.

SIZE 16 mm.

HABITAT Old woodland and blackthorn hedges and scrub.

FOODPLANT Blackthorn.

RANGE Most of Europe except far N and S; in Britain only in English Midlands.

FLIGHT June–August.

SIMILAR SPECIES Ilex Hairstreak (p.106) and White-letter Hair-streak (p.107) are similar below, but have much smaller black spots.

caterpillar

Named for the conspicuous blue spot at the rear of the hindwing, this butterfly rarely opens its wings at rest. The upperside is dull brown, sometimes with an orange flush in the female. Winter is passed as an egg.

SIZE 16 mm.

HABITAT Dry scrub, especially in hilly areas.

FOODPLANT Blackthorn, buckthorn and other shrubs.

RANGE S & C Europe except British Isles.

FLIGHT June–July.

SIMILAR SPECIES No other hairstreak has such a large blue spot on hindwing.

caterpillar

Ilex Hairstreak

Nordmannia ilicis

The upperside, rarely seen at rest, is dull brown, with a variable orange patch on the female forewing. Look for the small black spots on each side of the orange spots on the underside of the hindwing. Winter is passed as an egg.

SIZE 18 mm.

HABITAT Light woodland and scrub.

FOODPLANT Oaks, especially the bushy kermes oak.

RANGE S & C Europe, including S Sweden, but not British Isles.

FLIGHT June–August.

SIMILAR SPECIES Sloe Hairstreak and False Ilex Hairstreak have paler underside with very faint black spots.

This butterfly is named for the W-shaped white line on the underside of the hindwing, although the W is not always clearly formed. The sooty brown upperside is never revealed at rest. The butterfly is very fond of bramble blossom. Winter is passed as an egg.

SIZE 16 mm.

HABITAT Light woodland, especially at edges, and old hedgerows.

FOODPLANT Elms.

RANGE Most of Europe, except Scotland, Ireland and far N.

FLIGHT June–August.

SIMILAR SPECIES Black Hairstreak (p.104) has darker underside with more extensive orange border. Ilex Hairstreak (opposite) has less orange.

caterpillar

This lively species commonly basks on or near the ground with its wings wide open, displaying gleaming coppery forewings. Look for the broad margin and irregular line of black spots in the outer part of the forewing. The forewing underside is a paler version of the upperside. The underside of the hindwing is greyish brown, with black dots and a red marginal band. Winter is passed as a caterpillar.

SIZE 14 mm.

HABITAT Rough grassland, wasteland and heathland.

FOODPLANT Sorrels and occasionally docks.

RANGE All Europe.

FLIGHT February–November, in two or three broods.

SIMILAR SPECIES Female Sooty Copper (p.111) has similar upperside, but outer spots in straight line.

caterpillar

The male, shown here, has a brilliant coppery upperside. The female is duller, with a row of elongated black spots on the forewing, often with largely brown hindwings. The underside has an orange forewing and a silvery-grey hindwing in both sexes, all spotted with black. Look for the grey outer margin of the forewing. Winter is passed as a caterpillar.

SIZE 20 mm.

HABITAT Fens, marshes and damp grassland.

FOODPLANT Great water dock and some other docks.

RANGE Scattered through C & SE Europe; endangered due to land-drainage.

FLIGHT May–September, in one or two broods.

SIMILAR SPECIES
Scarce Copper (p.110) and Purple-shot Copper (p.112) lack grey margin under forewing.

caterpillar

pupa

The male, shown above, has a brilliant upperside with narrow black margins and usually without black spots on the fore-wing. The female, shown below, is much duller and heavily spotted with black, and her hindwings are often brown with an orange border. Look for the greenish yellow underside, the hindwing spotted with white and with a faint red border. Winter is passed as an egg.

SIZE 17 mm.

HABITAT Flower-rich grassland, especially near water.

FOODPLANT Docks and sorrels.

RANGE Widespread, but absent from Britain, adjacent parts of NW Europe and far N.

FLIGHT June–August.

SIMILAR SPECIES No other copper has greenish yellow underside with white spots.

♀

The male upperside is sooty brown, with or without an orange border, and with faintly visible black spots. The female forewing is orange, with a fairly straight row of dark spots in the outer half. Her hindwing is brown with an orange border. The underside is greenish grey with black spots and orange borders in both sexes, but the female may have an orange flush on the forewing. Winter is passed as a caterpillar.

SIZE 15 mm.

HABITAT Flower-rich grassland and rough ground.

FOODPLANT Various docks.

RANGE S & C Europe except British Isles.

FLIGHT April–October, in two broods.

SIMILAR SPECIES Small Copper (p.108) is like female, but is generally brown under the hindwing.

Purple-Shot Copper

Heodes alciphron

Males have a strong violet sheen over most of the upperside, although this is much reduced in the Alps and south-west Europe. Females are either brown, with an orange border to the hindwing, or bright orange. Both sexes have an irregular line of black spots in the outer part of the forewing. The underside of both sexes is greyish yellow with black spots and orange-bordered hindwings. Winter is passed as a caterpillar.

♀ ▲

SIZE 18 mm.

HABITAT Flower-rich fields and hillsides.

FOODPLANT Various docks.

RANGE S & C Europe except British Isles.

FLIGHT June–August.

SIMILAR SPECIES Violet Copper is much smaller. The female Large Copper (p.109) is silvery grey below.

The male's fiery red upperside is generally tinged with purple on the front edge of the forewing and on the rear half of the hindwing, but is otherwise almost unmarked. The female has a duller forewing, with an even row of black spots, and generally has a brown hindwing with an orange border. The underside is yellowish grey with black spots and a variable orange flush on the forewing. Winter is passed as a caterpillar.

SIZE 17 mm.

HABITAT Damp grassland.

FOODPLANT Various docks and bistorts.

RANGE All Europe, except Britain and S Iberia.

FLIGHT June–August.

SIMILAR SPECIES None.

♀

Lampides boeticus

Named for the slender 'tail' on the hindwing. The upperside, rarely seen at rest, is violet blue and rather hairy in the male, shown here, but sooty brown with a blue basal flush in the female. The underside is brown, with a broad white band on the hindwing. Passes winter as a pupa.

SIZE 18 mm.

HABITAT Rough, flowery places; sometimes gardens.

FOODPLANT Gorse, lupins and other large legumes; feeds in seed pods.

RANGE Resident in S Europe; migrates northwards in spring, but rarely seen north of the Alps.

FLIGHT May–October, in two or three broods.

SIMILAR SPECIES Lang's Short-tailed Blue is smaller, with more obvious stripes on underside.

The upperside, not often seen at rest, is violet blue in the male and sooty brown in the female, sometimes with a dusting of blue at the base. Look for the short 'tail' on the hindwing and for the two neighbouring orange spots on the bluish grey underside. Winter is passed as an egg or caterpillar.

SIZE 14 mm.

HABITAT Flower-rich grassland, mainly in damp places.

FOODPLANT Trefoils and other herbaceous and shrubby legumes, including gorse.

RANGE S & C Europe; a scarce vagrant to Britain, where it was once known as the Bloxworth Blue.

FLIGHT April–October, in two or three broods.

SIMILAR SPECIES Several are superficially similar, but all lack orange spots and 'tail'.

♀

Cupido minimus

The male upperside is sooty brown with a dusting of blue scales at the base. The female is similar, but has no blue. The silvery-grey underside carries small black spots, those on the forewing forming an almost straight line. Winter is passed as a fully-grown caterpillar. This is the smallest British butterfly.

SIZE 12 mm.

HABITAT Rough grassland, mainly on chalk and limestone.

FOODPLANT Kidney vetch.

RANGE Most of Europe except S Iberia.

FLIGHT April–September, in one or two broods.

caterpillar

SIMILAR SPECIES Female Lorquin's Blue is almost identical, but lives only in S Iberia, where Small Blue is absent. Several other female blues are similar, but larger or with 'tails'.

The upperside is pale violet blue, with a narrow black border in the male and a broader one in the female. The border is especially broad on the forewing of the summer brood. The underside is powdery blue with distinctly elongated black spots. Winter is passed as a pupa.

SIZE 16 mm.

HABITAT Open woodland, hedgerows, parks and gardens.

FOODPLANT Mainly holly in spring and ivy in autumn, but also several other shrubs, especially in spring.

RANGE Almost all Europe except Scotland.

FLIGHT April–September, in two broods.

SIMILAR SPECIES Several have similar underside, but are greyer or have 'tails'.

♀

Green-Underside Blue

Glaucopsyche alexis

Named for the blue-green flush at the base of the underside of the hindwing. Notice also the large black spots under the forewing and the lack of any marginal spots. The upperside is violet blue with a dark border in the male, and dark brown with a faint blue dusting at the base in the female. Winter is passed as a caterpillar.

SIZE 16 mm.

HABITAT Flowery grassland and woodland edges, mainly in uplands.

FOODPLANT Many leguminous herbs and shrubs.

RANGE Most of Europe, except Britain and far N.

FLIGHT April–July.

SIMILAR SPECIES Mazarine Blue (p.125) has blue flush below, but hindwing is browner and has more spots.

The female upperside is heavily dusted with black and has a wider black border than the male, shown here. The underside of both sexes is greyish brown with large black spots, and a variable blue flush at the base of the hindwing. Winter is passed as a caterpillar, in a red ant nest. The species is endangered.

SIZE 20 mm.

HABITAT Rough grassland with red ant populations.

FOODPLANT Wild thyme in the early stages; older caterpillars feed on ant grubs.

RANGE Most of Europe except far N; extinct in Britain by 1979, but now re-established.

FLIGHT June–July.

SIMILAR SPECIES Scarce Large Blue has smaller spots, and no blue flush on underside.

Chequered Blue

Scolitantides orion

The upperside is sooty black in both sexes, with a basal blue flush, usually better developed in the male, seen here, than in the female although both sexes are quite blue in Scandinavia. The underside is almost white with large black spots, and an orange band on the hindwing only. Look for the strongly chequered margins on both upper and lower surfaces. Winter is passed as a caterpillar.

SIZE 14 mm.

HABITAT Dry, stony hillsides and coastal cliffs.

FOODPLANT Stonecrops.

RANGE S Europe and S Scandinavia.

FLIGHT April–August, in one or two broods.

SIMILAR SPECIES None.

The male upperside, shown here, is deep blue with conspicuous black borders and white fringes. The female upperside is dark brown, often tinged with blue at the base and often with orange spots around the edges. The underside is silvery grey in the male and brown in the female, but both have tiny shiny blue spots – the silver studs – near the edge of the hindwing. Winter is passed as an egg.

SIZE 14 mm.

HABITAT Mainly heathland, but also rough grassland.

FOODPLANT Gorse, other legumes and heathers.

RANGE All Europe, except the Arctic, Scotland and Ireland.

FLIGHT May–September, in one or two broods.

SIMILAR SPECIES None in Britain. Idas Blue on the Continent is almost identical.

♀

Cranberry Blue

Vacciniina optilete

The upperside is deep violet blue in the male and dark brown with a violet tinge in the female. The underside is grey with large black spots, but look also for the one or two red and blue spots near the rear edge of the hindwing. Winter is passed as a caterpillar.

SIZE 15 mm.

HABITAT Moors and damp grassland; mainly in mountains, but at low levels in N.

FOODPLANT Cranberry, other *Vaccinium* species and various heathers.

RANGE N & C Europe, except France and British Isles; a rapidly declining species.

FLIGHT July.

SIMILAR SPECIES None.

caterpillar

The upperside is chocolate brown in both sexes, with no trace of blue. All wings are bordered with orange spots, which are usually smaller in the male than in the female, seen here. The underside is greyish brown with red spots around the margins and a white flash on the hindwing. The two black spots near the front edge of the hindwing lie one above the other to form a colon-shape. Winter is passed as a caterpillar.

SIZE 14 mm.

HABITAT Heaths and rough, grassy places.

FOODPLANT Rock-roses and storksbills.

RANGE S & C Europe, except Scotland and Ireland.

FLIGHT April–September, in two or three broods.

SIMILAR SPECIES Mountain Argus (p.124) has fewer orange spots on upperside.

caterpillar

Aricia artaxerxes

The upperside is chocolate brown, usually with small red marginal spots. The white spot on the forewing, characteristic of most British colonies, is often absent elsewhere. The underside commonly has white spots in the British race, but is otherwise like the Brown Argus. Passes winter as a caterpillar. Also called the Northern Brown Argus.

♀

SIZE 13 mm.

HABITAT Rough grassland, usually in mountains.

FOODPLANT Rock-roses and storksbills.

RANGE Scattered over most of Europe, except Ireland and S Britain.

FLIGHT June–August.

SIMILAR SPECIES Brown Argus (p.123) usually has more red on upperside and never has white forewing spot.

Cyaniris semiargus

The upperside of the male is deep violet blue with narrow black margins, while the female is chocolate brown with a faint violet tinge at the base. The underside is pale brown with black spots and no marginal markings. Winter is passed as a caterpillar.

SIZE 16 mm.

HABITAT Flowery grassland; males often gather in large numbers to drink from muddy ground.

FOODPLANT Clovers and other low-growing legumes.

RANGE Most of Europe, except far N and British Isles; although vagrants sometimes reach England.

FLIGHT May–August, in one or two broods.

SIMILAR SPECIES Greek Mazarine Blue has red zigzag on underside of hindwing.

caterpillar

Damon Blue

Agrodiaetus damon

The male upperside is pale blue with wide sooty brown borders and plain white fringes. The female is brown with no more than a few blue scales at the base. The underside is greyish brown in both sexes, with large black spots on the forewing. Look for the conspicuous white stripe under the hindwing. There are no marginal markings. Winter is passed as a small caterpillar.

SIZE 17 mm.

HABITAT Flowery mountain slopes, usually on limestone.

caterpillar

FOODPLANT Sainfoins.

RANGE S & C Europe except British Isles.

FLIGHT July–August.

SIMILAR SPECIES Furry Blue has paler underside with less obvious white streak. Chalkhill Blue (p.128) has chequered fringes.

Look for the scalloped edge at the rear of the hindwing to separate this species from all other blues. The upperside is shining blue in both sexes, but the female has wider black margins than the male, shown here. The underside is pale grey with black spots and very faint marginal markings. Winter is passed as an egg or a small caterpillar.

SIZE 18 mm.

HABITAT Warm, flower-rich hillsides.

FOODPLANT Wild thyme and various low-growing legumes.

RANGE S Europe, extending northwards to Hungary in E.

FLIGHT May–August.

SIMILAR SPECIES None.

Chalkhill Blue

Lysandra coridon

The male upperside is silvery blue, with a wide sooty brown margin on the forewing and conspicuous black dots around the edge of the hindwing. The female is brown with faint orange spots on the edge of the hindwing. The fringes are chequered in both sexes. The underside is pale grey with black spots and faint orange markings. Winter is passed as an egg.

♀

SIZE 18 mm.

HABITAT Flowery slopes on chalk and limestone.

FOODPLANT Horseshoe vetch and other small legumes.

RANGE S & C Europe, as far as S England.

FLIGHT June–August.

SIMILAR SPECIES Provence Chalkhill Blue flies at a different time in S Europe. Female Adonis Blue (opposite) has blue scales at edge of hindwing.

The male upperside is a brilliant sky blue. The female is chocolate brown with a few blue scales near the base, and with orange spots, bordered by blue scales, around the edge of the hindwing. Look for the chequered fringes in both sexes. The underside is brownish grey with black and orange spots. Winter is passed as a small caterpillar.

SIZE 17 mm.

HABITAT Flowery hillsides on chalk and limestone.

FOODPLANT Horseshoe vetch.

RANGE S & C Europe, as far as S England.

FLIGHT May–September, in two broods.

♀

SIMILAR SPECIES Female Chalkhill Blue (opposite) is usually paler, with white scales beyond marginal spots on upperside of hindwing.

Common Blue

Polyommatus icarus

The male upperside, seen here, is violet blue with narrow black margins and no chequered fringes. The female is brown, often heavily dusted with blue and with a variable amount of orange spotting on the margins. The underside is brownish grey with black and orange spots. Look for the small spot in the centre of the forewing cell – no similar blue has this spot. Winter is passed as a small caterpillar.

SIZE 17 mm.

HABITAT Grassy places of all kinds.

FOODPLANT Birdsfoot trefoil and other legumes.

RANGE All Europe; one of Europe's commonest butterflies.

FLIGHT April–October, in 1–3 broods.

SIMILAR SPECIES Several continental species are superficially similar, but most lack the spot in the cell.

The male, shown here, is silky white above and dull brown below, and dances in a ghost-like manner over the vegetation to attract females at dusk. Females have a yellowish brown upperside. Two winters are passed as a caterpillar, which can cause severe damage to cereals and to carrots and other root crops.

SIZE 22 mm.

HABITAT Grassy places of all kinds, including gardens, orchards and wasteland.

FOODPLANT Roots of grasses and other herbaceous plants, also roots of young trees.

RANGE N & C Europe.

FLIGHT June–August.

SIMILAR SPECIES None.

♀

A sturdy moth named for the strong goat-like smell of its caterpillar. All four wings are dull greyish brown and the abdomen is clearly ringed with black and grey. At rest, with its wings wrapped around the body, the moth looks very like a piece of twig. The wood-boring caterpillar may overwinter three or four times before maturing.

SIZE 35–45 mm.

HABITAT Almost anywhere with deciduous trees.

FOODPLANT The larva tunnels in trunks of a wide range of deciduous trees.

RANGE All Europe.

FLIGHT June–July.

caterpillar   **SIMILAR SPECIES** None.

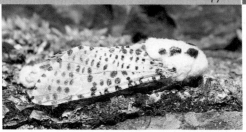

A rather hairy moth with thinly scaled wings, all four with black spots. At rest, the tip of the abdomen protrudes beyond the wings. That of the male bears a white tuft of hair, while the female abdomen ends in a conspicuous ovipositor. The wood-boring caterpillar overwinters at least twice before maturing.

SIZE 20–35 mm.

caterpillar

HABITAT Almost anywhere with deciduous trees; not uncommon in parks and orchards.

FOODPLANT The larva tunnels in trunks of a wide range of deciduous trees.

RANGE Most of Europe, except Scotland and Ireland.

FLIGHT June–August.

SIMILAR SPECIES Puss Moth (p.193) is similar at rest, but stouter and with zig-zag patterns on its wings.

Six-Spot Burnet

Zygaena filipendulae

A sluggish, day-flying moth, reluctant to take to the air. It usually has six clear red spots on the forewing, although the two outer spots may merge into one. The hindwing is red with a narrow black border. The moth is most often seen feeding at scabious and knapweed flowers. Winter is passed as a caterpillar.

SIZE 11–18 mm.

HABITAT Grassy places of all kinds.

FOODPLANT Birdsfoot trefoil and other low-growing legumes.

RANGE Almost all Europe.

FLIGHT June–August.

SIMILAR SPECIES
Several, mostly with fewer
spots and/or broader
hindwing margins.
Some have a red belt
on abdomen.

**Six-spot Burnet
exuding toxins**

A sluggish, day-flying moth with a red collar and usually with a red abdominal belt. The red or orange spots are always joined up, but may lack the white edges. The hindwing is red with a narrow black border. Winter is passed as a caterpillar.

SIZE 12 mm.

HABITAT Rough grassy places, usually on lime-rich soil.

FOODPLANT Crown vetch and other small legumes.

RANGE SW Europe, extending through Alps to Germany; most common in S.

FLIGHT June–August.

SIMILAR SPECIES *Z. carniolica* (p. 137) has pale collar and separate spots. *Z. hilaris* has pale collar and no red belt.

Variable Burnet

Zygaena ephialtes

A very variable moth, often red and black like the Six-spot Burnet (p.134) in Central Europe, but always with a red belt. The spots are mainly white and the hindwing is black in the south. In Italy and south-east Europe the red is replaced by yellow. Winter is passed as a caterpillar.

SIZE 15 mm.

HABITAT Rough grassland and open woodland.

FOODPLANT Crown vetch and related legumes.

RANGE S & C Europe except British Isles.

FLIGHT June–September.

SIMILAR SPECIES Yellow-and-white spotted form is like Nine-spotted Moth (p.206), which has several white spots on hindwing and does not have clubbed antennae.

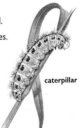

caterpillar

A sluggish, day-flying moth with pale-ringed spots, the outer one always crescent-shaped. There is a white collar and often a red abdominal belt. The hindwing is red with a black border. Winter is passed as a caterpillar.

SIZE 15 mm.

HABITAT Rough grassland and scrub, usually in uplands.

FOODPLANT Birdsfoot trefoil and other low-growing legumes.

RANGE S & C Europe, mainly in E.

FLIGHT June–August.

SIMILAR SPECIES Z. *occitanica* from SW Europe always has red belt and outer spot is usually pure white.

A sluggish, day-flying moth with forewings ranging from golden green to distinctly bluish. The hindwing is grey. The antenna gradually thickens towards the tip. Winter is passed as a caterpillar. There are two biologically distinct forms, which may turn out to be different species. One inhabits damp grassland and the other favours dry habitats.

SIZE 12 mm.

HABITAT Rough grassland and woodland clearings.

FOODPLANT Common sorrel.

RANGE Most of Europe except far N.

FLIGHT May–August.

SIMILAR SPECIES Several, differing mainly in size and in structure of antennae.

male antenna

The moth is bright green when fresh but, as in all emeralds, the colour soon fades and the lines of white spots disappear. Older specimens are very pale and some appear almost white. Adults are commonly attracted to lights at night. Winter is passed as a caterpillar.

SIZE 30 mm.

HABITAT Woodland, hedgerows, scrub and heathland.

FOODPLANT Assorted trees, mainly birch and hazel.

RANGE All Europe.

FLIGHT June–August.

SIMILAR SPECIES None of the other emeralds approaches this size.

caterpillar

Look for the pale blotches at the rear of each wing to distinguish this from all other emerald moths, although they are not quite so easy to see in older specimens when the green has faded. Winter is passed as a small caterpillar.

SIZE 15 mm.

HABITAT In and around oakwoods.

FOODPLANT Oak.

RANGE C Europe, as far as S Scandinavia, but only in S half of Britain.

FLIGHT June–July.

SIMILAR SPECIES None.

Sessile Oak

Look for the red legs, red-chequered wing margins, and the shallow notch between two prominent points on the outer edge of the hindwing. The wings are bright green when fresh, but soon fade to a dull bluish green. The male antennae are strongly feathered. Winter is passed as a caterpillar.

SIZE 17 mm.

HABITAT Light woodland and scrub; mainly coastal in British Isles.

FOODPLANT Yarrow, gorse and many other plants.

RANGE Most of Europe except far N, but confined to S coast of Britain

FLIGHT June–August.

SIMILAR SPECIES Common Emerald has single point and no obvious notch in hindwing.

Yarrow

Timandra griseata

Named for the prominent pink or purple 'vein' running across each wing. The ground colour of the wings varies from cream to pale grey. Look also for the purplish wing margins and the prominent point on the hindwing.

SIZE 16 mm.

HABITAT Hedgerows and other rough places.

FOODPLANT Docks, sorrels, knotgrasses and many other low-growing plants.

RANGE Most of Europe except far N.

FLIGHT May–September, in two broods.

SIMILAR SPECIES
Small Bloodvein and Maiden's Blush are both yellower and 'vein' runs from front edge of the forewing, not from tip.

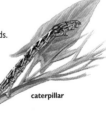

caterpillar

The delicate, lace-like pattern around the edges of all four wings varies in density, but the moth is instantly recognizable. It frequently comes to lights at night. Winter is passed as a caterpillar.

SIZE 12 mm.

HABITAT Rough grassland, mainly on lime-rich soils.

FOODPLANT Marjoram and wild thyme.

Thyme

RANGE Most of Europe except far N, but confined to S Britain, where it appears to be declining.

FLIGHT May–September, in one or two broods.

SIMILAR SPECIES Treble Brown Spot is smaller and browner, with lacy pattern only on forewing.

Marjoram

Garden Carpet

Xanthorhoe fluctuata

Rather variable, but look for the black 'shoulders' and the notched rectangular patch at the front of the forewing. The hindwing is densely lined and speckled with grey. The triangular shape of the resting moth, with the hindwings completely hidden, is typical of the carpet moths which are named for their striped and mottled patterns. Winter is passed as a pupa.

SIZE 13 mm.

HABITAT Gardens and wasteland; common in towns.

FOODPLANT Many wild and cultivated crucifers.

RANGE All Europe.

FLIGHT April–October, in two or three broods.

SIMILAR SPECIES
Common Carpet has black band right across forewing. Many others are superficially similar, but have white hindwings.

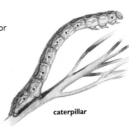

caterpillar

The ground colour of the wings ranges from pale yellow to rich brown, with dark specimens most common in northern areas. The delicate lines, resembling those on a seashell, are always visible, and the dark band across the forewing is sometimes much broader than shown here. Winter is passed as a caterpillar.

SIZE 14 mm.

HABITAT Hedgerows, gardens, fields and wasteland.

FOODPLANT Grasses, docks and many other low-growing plants.

RANGE All Europe except far N.

FLIGHT June–August.

SIMILAR SPECIES None.

caterpillar

Cidaria fulvata

Like many of its relatives, the moth commonly rests with its abdomen sticking up perpendicular to the wings. The hindwing, rarely seen when the insect is at rest, is cream with a yellow border. Winter is passed as an egg.

SIZE 14 mm.

HABITAT Gardens, hedgerows and wasteland.

FOODPLANT Wild and cultivated roses.

RANGE All Europe except far N.

FLIGHT May–July.

SIMILAR SPECIES Northern Spinach is larger with paler central band, and holds its wings well away from its body.

caterpillar

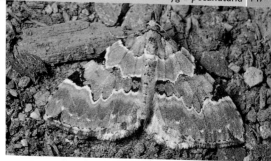

The 'shoulders' and the central band of the forewing have a slight yellow tint, contrasting with the more bluish green of the rest of the wing in freshly emerged insects, but the whole wing soon fades to a dull yellowish green. The hindwing is grey. Winter is passed as a caterpillar.

SIZE 13 mm.

HABITAT Hedgerows, heaths and other rough places.

FOODPLANT Bedstraws.

RANGE All Europe.

FLIGHT April–September, in one or two broods.

SIMILAR SPECIES Beech-green Carpet has a green ground colour, but central band is distinctly brownish.

Lady's Bedstraw

When the moth is resting on tree-trunks, with the front edges of its forewings in a straight line, the dark, wavy bands on the wings break up its outline very effectively and help it to merge with the bark patterns. Winter is passed as a pupa, beautifully camouflaged on the foodplant.

caterpillar

SIZE 16 mm.

HABITAT Hedgerows, scrub and woodland edges.

FOODPLANT Traveller's joy.

RANGE Most of Europe except far N.

FLIGHT May–September, in two broods.

SIMILAR SPECIES Waved Umber (p.163) has scalloped hindwing and dark band on forewing is incomplete.

Only the male, shown here, has proper wings, which range from pale grey to sooty brown. The female has no more than tiny stumps and sits on tree-trunks and branches waiting for a mate to arrive. The male commonly comes to lighted windows on winter evenings and may sit there for hours.

caterpillar

SIZE 15 mm (male).

HABITAT Anywhere with deciduous trees, including gardens, parks and orchards.

FOODPLANT Almost any deciduous tree; a serious pest of apples and pears.

RANGE All Europe, including Iceland.

FLIGHT October–February.

SIMILAR SPECIES Northern Winter Moth is larger, with whiter hindwings.

Eupithecia centaureata

The ground colour is white or grey and the dark blotch on the forewing may be very small. The resting moth is easily mistaken for a bird dropping. The resting position, shown here, with the forewings more or less in a straight line at right angles to the body and only a small part of the hindwings exposed, is typical of the pug moths. Winter is passed as a pupa.

SIZE 11 mm.

HABITAT Almost anywhere; common in gardens.

FOODPLANT Flowers of many herbaceous plants, especially yarrow and other composites.

RANGE All Europe.

FLIGHT May–October, in one or two broods.

SIMILAR SPECIES None.

caterpillar

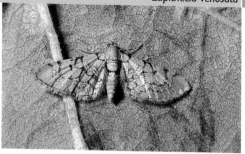

The ground colour of the forewing ranges from almost white, through dull grey, to mid-brown, and the intensity of the black and white bands also varies, but the outer white band is usually clearly visible. Winter is passed as a pupa.

SIZE 11 mm.

HABITAT Rough grassland, including roadsides and coastal cliffs, usually on lime-rich soils.

FOODPLANT Seed capsules of bladder campion and sea campion.

RANGE All Europe.

FLIGHT May–July.

SIMILAR SPECIES None.

caterpillar

Magpie Moth

Abraxas grossulariata

A fragile, weak-flying moth with ground colour ranging from white to butter yellow. The wing pattern is also very variable, but the wavy yellow stripe is usually visible on the forewing. Look also for the black-spotted yellow abdomen. Winter is passed as a caterpillar, which has the same colours as the adult and is often a pest of gooseberries and currants.

SIZE 22 mm.

HABITAT Hedgerows and gardens.

FOODPLANT Blackthorn and many other shrubs, including currants and gooseberry.

RANGE All Europe except far N.

FLIGHT June–August.

SIMILAR SPECIES None.

caterpillar

A rather weak, day-flying moth, usually taking to the air only in sunshine. The brown markings vary greatly in size, some specimens appearing almost completely yellow and others almost brown. Winter is passed as a pupa.

SIZE 14 mm.

HABITAT Woodland edges, hedgerows and scrub.

FOODPLANT Wood sage, yellow deadnettle and other labiates.

RANGE All Europe.

FLIGHT May–June.

SIMILAR SPECIES None.

This very common moth is often mistaken for a butterfly because of its bright colour and the fact that it often flutters around by day if it is disturbed. It commonly settles on lighted windows at night. Winter is passed either as a caterpillar or a pupa.

SIZE 15–22 mm.

HABITAT Light woodland, hedgerows and scrub.

FOODPLANT Hawthorn, blackthorn and other rosaceous shrubs and trees, including orchard plums.

RANGE All Europe.

FLIGHT April–October, in one or two broods.

SIMILAR SPECIES None.

caterpillar

Also known as the Lilac Thorn, this attractive moth rests in a most unusual position, with its forewings crumpled and seemingly distorted – with the result that the moth is often mistaken for a shrivelled leaf. The male, shown here, is more brightly coloured than the female. Winter is passed as a small caterpillar.

SIZE 20 mm.

HABITAT Hedgerows, gardens and light woodland.

FOODPLANT Honeysuckle, lilac, privet and ash.

RANGE All Europe, except Scotland and far N.

FLIGHT May–September, in one or two broods.

SIMILAR SPECIES None.

caterpillar

Look for the bright canary-yellow hairs on the thorax. The wings are basically yellow, but often heavily dusted with brown scales, as shown here. The moth commonly rests with its wings partly raised to form a shallow V. Winter is passed as an egg.

caterpillar

SIZE 18 mm.

HABITAT Light woodland, hedgerows, gardens, fens and heathland.

FOODPLANT Birch, sallow, alder and various other deciduous trees.

RANGE All Europe.

FLIGHT June–October.

SIMILAR SPECIES September Thorn has clearer yellow wings with no dark spot between cross lines. August Thorn has sharply bent inner cross line.

Moths of the spring brood are larger and paler than the summer brood, shown here. Females in each brood are usually paler than the males. The moths rest with their wings closed over the body. The upperside resembles the underside in colour, but the hindwing is largely unmarked. Winter is passed as a pupa.

SIZE 18–24 mm.

HABITAT Almost anywhere with deciduous trees or shrubs.

caterpillar

FOODPLANT Hawthorn, blackthorn and many other deciduous species.

RANGE All Europe.

FLIGHT April–September, in one or two broods.

SIMILAR SPECIES
Purple Thorn is much darker. Lunar Thorn has very jagged hindwings.

The forewing ground colour ranges from cream to brick-red. The central band is usually noticeably darker than the rest of the wing, but it and the enclosing cross lines are occasionally indistinct. The dark spot is always present. The hindwing is cream or buff. Winter is passed as an egg.

SIZE 22 mm.

HABITAT Almost anywhere with trees; common in parks and gardens.

FOODPLANT A wide range of deciduous trees and shrubs.

RANGE All Europe.

FLIGHT June–August.

SIMILAR SPECIES Feathered Thorn is usually much browner, lacking dark spot on forewing.

caterpillar

An unmistakable moth named for the little 'tail' on each hindwing. The wings are bright lemon yellow at first, but soon fade to pale cream. Its lazy flight, commencing at dusk, gives the moth a ghost-like appearance. It often comes to lighted windows. Winter is passed as a caterpillar.

SIZE 30 mm.

HABITAT Woods, hedgerows and gardens.

FOODPLANT Ivy, hawthorn and many other shrubs.

RANGE All Europe except far N.

FLIGHT June–August.

SIMILAR SPECIES None.

caterpillar

Oak Beauty

Biston strataria

The white areas of the forewing are sometimes heavily speckled with sooty brown, making the moth appear almost completely brown. The hindwing is essentially white, but always heavily dusted with brown. Winter is passed as a pupa.

SIZE 26 mm.

HABITAT Woods and parkland with plenty of trees.

caterpillar

FOODPLANT Oak, elm, alder, birch and many other deciduous trees.

RANGE Most of Europe except far N.

FLIGHT February–May.

SIMILAR SPECIES Heavily speckled specimens may resemble Peppered Moth (opposite).

The typical moth, shown here, is white with a dense peppering of black, but the species also occurs in a black or melanic form, with just a small white spot on each 'shoulder'. The melanic form spread rapidly during the late 19th and early 20th centuries as a result of air pollution, which blackened trees and buildings and gave the black moths a better chance of concealment. Winter is passed as a pupa.

SIZE 22–28 mm.

HABITAT Woods, hedges and gardens; common in towns.

FOODPLANT A wide range of deciduous trees and shrubs.

RANGE All Europe except far N.

FLIGHT May–August.

SIMILAR SPECIES Dark forms of Oak Beauty (opposite) may be similar.

caterpillar

Only the male is winged; the female, looking more like a spider, sits on tree-trunks and branches waiting for a mate. The male's forewing ranges from white with brown markings to brick-red with almost black markings, and is sometimes uniformly brown with darker speckles. The hindwing is white with brown speckles. Pupa, adult and eggs can all be found in the winter.

SIZE 20 mm.

HABITAT Anywhere with trees; common in gardens.

FOODPLANT Almost any deciduous tree or shrub.

RANGE All Europe, including Iceland.

FLIGHT September–March, but mainly October–December.

SIMILAR SPECIES Dotted Border has dots around edge of hindwing.

caterpillar

When resting on tree-trunks in its normal position, with its body horizontal, the wing pattern blends beautifully with the bark fissures and the moth is extremely difficult to see. The ground colour ranges from pale buff to mid-brown and tends to be darker in males. Melanic individuals are not uncommon in urban areas. Winter is passed as a pupa.

SIZE 19 mm.

HABITAT Light woodland, hedgerows, parks and gardens.

FOODPLANT Privet, lilac, ash and various other trees.

RANGE S & C Europe.

FLIGHT April–June.

SIMILAR SPECIES
Small Waved Umber
(p.148) has dark band
right across forewing,
and edge of hindwing is
not scalloped.

caterpillar

Light Emerald

Campaea margaritata

This is not closely related to the other emerald moths. The cross lines are almost straight and each consists of a dark line edged with white. The freshly emerged moth is a delicate green, but the colour quickly fades to greyish or even white. Winter is passed as a caterpillar.

SIZE 20–25 mm.

HABITAT
Light woodland, parks
and hedgerows.

FOODPLANT Oak, birch,
hawthorn and many other
deciduous trees.

caterpillar

RANGE All Europe.

FLIGHT May–September, in one or two broods.

SIMILAR SPECIES Other emeralds all have plain white and rather wavy cross lines.

Named for the hooked wing-tips and the pebble-like spot on the forewing. The forewing, usually somewhat darker than the hindwing, ranges from off-white to tawny and the resting moth resembles a dead leaf. The male has feathery antennae. Winter is passed as a pupa.

SIZE 35–40 mm.

HABITAT Heaths and light woodland.

FOODPLANT Birch and alder.

RANGE All Europe.

FLIGHT April–September, in two broods.

SIMILAR SPECIES
Dusky Hooktip has a similar
colour, but no 'pebble'.

caterpillar

Named for the delicate pink patches on the forewing, although these patches are occasionally pale brown. The hindwing is silky grey. The moth frequently comes to moth traps and lighted windows at night. Winter is passed as a pupa.

SIZE 20 mm.

HABITAT Deciduous woodland and scrub.

FOODPLANT Bramble.

RANGE All Europe.

FLIGHT May–September, in one or two broods.

SIMILAR SPECIES None.

caterpillar

A conspicuous white stripe divides the forewing into a greyish basal area and a browner outer region. Look for the distinctive 'squiggles' – the arches – in the outer region. The hindwing is greyish brown. The moth often comes to lights at night. Winter is passed as a pupa.

SIZE 20 mm.

HABITAT Deciduous woodland, hedgerows and scrub.

FOODPLANT Bramble and raspberry.

RANGE All Europe.

FLIGHT June–August, in one or two broods.

SIMILAR SPECIES None.

caterpillar

Lackey Moth

Malacosoma neustria

The male, shown here, flies fast on rather short wings. The female is larger and has much longer wings. Both sexes range from pale yellow with dark cross lines to reddish brown with pale cross lines, and the area between the lines is often much darker than the rest of the wing. Winter is passed as an egg.

SIZE 14–20 mm.

HABITAT Woodlands, parks, gardens, orchards and hedgerows; often common in towns.

FOODPLANT Many deciduous trees; often an orchard pest.

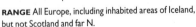

♀

RANGE All Europe, including inhabited areas of Iceland, but not Scotland and far N.

FLIGHT June–August.

SIMILAR SPECIES Ground Lackey has inner cross line curving in to meet base of wing.

The moth shown here is a male. The female is larger and yellowish brown all over, although the basal area of the wing is slightly darker than the rest. The hindwing is similar, but lacks white spot. Males fly rapidly by day, but females fly at dusk, scattering their eggs as they fly slowly over the vegetation. Winter is passed as a small caterpillar.

SIZE 25–45 mm.

HABITAT Open woodland, scrub, heaths and moors.

FOODPLANT Bramble, blackthorn, heather and many other trees and shrub.

RANGE All Europe.

FLIGHT May–September.

SIMILAR SPECIES Grass Eggar is usually paler, with plain hindwings.

Macrothylacia rubi

The female, on the left, is much greyer than the male, on the right, and also has longer wings. The hindwing is the same colour, but with no markings. The male is active by day, but the female flies and lays her eggs at night. Winter is passed as a caterpillar.

SIZE 25–35 mm.

HABITAT Heaths, moors, grassland and open woodland.

FOODPLANT Heather, bilberry, bramble and many other low-growing plants.

RANGE All Europe.

FLIGHT May–July.

SIMILAR SPECIES Grass Eggar usually has only one cross line and always has a white spot on forewing.

caterpillar

The female is much yellower than the male, shown here, but both sexes can be distinguished from related moths by looking for the thin dark line running obliquely back from the wing-tip. The moth is named for the larval habit of drinking dew and raindrops. Winter is passed as a caterpillar.

SIZE 22–34 mm.

HABITAT Rough grassland, including woodland rides, roadside verges, hedgerows and fenland.

FOODPLANT Cocksfoot and other coarse grasses, including reeds.

RANGE Most of Europe except far N.

FLIGHT June–August.

SIMILAR SPECIES *Odonestis pruni* of C Europe has cross line from front edge, not wing-tip.

Gastropacha quercifolia

The serrated edges of the wings distinguish this from most other moths. At rest, with the hindwings more or less flat and the forewings held roofwise above them, the moth looks just like a cluster of dead leaves. The wings are sometimes cream or buff, especially in southern Europe. Winter is passed as a caterpillar.

SIZE 25–45 mm.

HABITAT Open woodland, scrub, orchards and hedgerows.

FOODPLANT Hawthorn, blackthorn and many other trees.

RANGE Most of Europe, except Scotland, Ireland and much of N Europe.

caterpillar

FLIGHT May–August.

SIMILAR SPECIES *Gastropacha populifolia* of C Europe is much paler, with no obvious cross lines.

Europe's largest moth, often mistaken for a bat when flying at night. The sexes are alike except for the male's very feathery antennae. The hindwing is like the forewing, but light brown at the front instead of grey. The moth does not feed. Winter is passed as a pupa, wrapped up in a very tough fibrous cocoon.

caterpillar

SIZE 70 mm.

HABITAT Light woodland, orchards and open country.

FOODPLANT Ash, blackthorn and various fruit trees; sometimes an orchard pest.

RANGE S Europe, as far N as S Switzerland.

FLIGHT March–June; occasionally later.

SIMILAR SPECIES Emperor Moth (p.174) is much smaller, with orange hindwing in male.

Emperor Moth

Pavonia pavonia

The female, shown here, flies rather sluggishly at dusk to lay her eggs. The male is a little smaller and much browner and flies rapidly by day, searching for the female scent with his feathery antennae. The moths do not feed. Winter is passed as a pupa, in a tough cocoon on the foodplant.

SIZE 25–40 mm.

HABITAT Heathland, scrub and hedgerows.

FOODPLANT Heathers, bramble, blackthorn, meadowsweet and various other plants.

RANGE All Europe.

FLIGHT April–June.

SIMILAR SPECIES Giant Peacock Moth (p.173) is much larger.

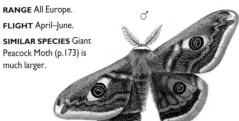

The ground colour ranges from yellow to deep brown and the eye-spots vary in size. The underside of the forewing also has an eye-spot. The male flies by day and the larger female is active at night, although it does not fly much. Resting moths may hold their wings vertically above the body like butterflies and resemble dead leaves. Winter is passed as a pupa.

SIZE 30–40 mm.

HABITAT Woodland, especially beechwoods.

FOODPLANT Beech, oak, birch and other trees.

RANGE Most of Europe, except British Isles and far N and S.

FLIGHT March–June.

SIMILAR SPECIES
None, although dark
specimens may resemble
Emperor Moth (opposite).

caterpillar

The female has shorter 'tails' than the male, shown here, and lacks the feathery antennae, but the moth is unmistakable. Eagerly sought by collectors, it is becoming rare and is legally protected in France. Winter is passed as a pupa.

SIZE 50 mm.

HABITAT Montane pinewoods.

FOODPLANT Various pines.

RANGE Mountains of Spain, and parts of French Alps (where probably introduced).

FLIGHT March–July.

SIMILAR SPECIES None.

caterpillar

Most often seen as a greyish blur, plunging its long tongue into petunias and other tubular flowers at dusk. The tongue is considerably longer than the body. Look for the grey thorax and pink-banded abdomen. The hindwing is grey and white. The moth arrives from Africa each spring and breeds in many areas, but does not survive the winter in Europe.

SIZE 45–55 mm.

HABITAT Anywhere with suitable flowers.

FOODPLANT Bindweeds and morning glory.

RANGE Summer visitor across Europe, including Iceland, but rare in Britain and N.

FLIGHT May–November.

caterpillar

SIMILAR SPECIES
Privet Hawkmoth (p.179) has black thorax and pink-striped hindwings.

Death's-Head Hawkmoth

Acherontia atropos

This sturdy moth is named for the skull-like pattern on its thorax. The hindwing is striped with yellow and black. The moth only has a short tongue and does not visit flowers, but often enters beehives to take honey. A great migrant, it arrives from Africa each year and breeds in some areas, but does not survive the winter in Europe.

SIZE 46–60 mm.

HABITAT Almost anywhere, but mostly near potato crops.

FOODPLANT Potatoes, nightshades and related plants.

RANGE Summer visitor across Europe; most common in S, rare in British Isles.

FLIGHT March–November; mainly late summer in Britain.

SIMILAR SPECIES None.

caterpillar

Unlike most hawkmoths, this species normally rests with its wings pulled tightly back along the body resembling a broken twig. Look for the pink-striped abdomen and hindwing. This is Britain's largest resident moth, but its population is reinforced each year by immigration from the Continent. Winter is passed as a pupa.

SIZE 45–55 mm.

HABITAT Woodland edges, scrub, parks and gardens.

FOODPLANT Privet, lilac and ash.

RANGE Most of Europe, except Scotland, Ireland and far N.

FLIGHT June–July.

SIMILAR SPECIES Convolvulus Hawkmoth (p.177) has grey thorax and no pink on hindwings.

caterpillar

Mimas tiliae

The forewing ground colour ranges from pale pink to olive green or brick-red, and with its 'ragged' wings swept back at rest, this moth is easily mistaken for a dead leaf. The central band, often broken in the middle and sometimes reduced to a single spot, is usually dark green. The hindwing is brownish. The moth does not feed. Winter is passed as a pupa.

SIZE 32–36 mm.

HABITAT Woods, parks and gardens; common in towns.

FOODPLANT Limes; less often on elm, birch and oak.

RANGE Most of Europe, except Scotland, Ireland and far N.

FLIGHT May–July.

SIMILAR SPECIES
Proserpinus proserpina (p.187) has similar pattern, but is much smaller.

caterpillar

The large eye-spot on the hindwing is normally concealed, but when the moth is disturbed it raises its forewings and displays the spots. At the same time it heaves its body up and down in a threatening manner, effectively scaring off birds and other predators. The moth does not feed. Winter is passed as a pupa.

SIZE 33–42 mm.

HABITAT Gardens, orchards and open woodland.

FOODPLANT Mainly sallows and apple, but will feed on several other deciduous trees.

RANGE All Europe, except Scotland and far N.

FLIGHT May–September, in one or two broods.

SIMILAR SPECIES None.

caterpillar

In its normal resting position, shown here, with the hindwings well forward of the forewings, the moth resembles a bunch of dead leaves. When disturbed, it raises its forewings and displays a brick-red patch on the hindwings, writhing about at the same time like the Eyed Hawkmoth (p.181). The adult does not feed. Winter is passed as a pupa.

SIZE 32–40 mm.

HABITAT Woodland edges, parks and gardens.

FOODPLANT Poplars and sallows.

RANGE All Europe, including the Arctic.

FLIGHT May–September, in one or two broods.

SIMILAR SPECIES None.

caterpillar

A rather slow-flying hawkmoth, easily recognized by its colour and the 'ragged' outer edges of the forewing. The hindwing is largely orange-brown, sometimes with a pinkish tinge. Winter is passed as a pupa.

SIZE 40–50 mm.

HABITAT Oak woodland.

FOODPLANT Oaks, especially cork oak and other evergreen species.

RANGE Mediterranean.

FLIGHT May–July.

SIMILAR SPECIES None.

caterpillar

Hemaris fuciformis

Most of the scales fall from the wings during the first flight, leaving them transparent apart from the chestnut-brown borders. The moth flies by day and hovers in front of a wide range of flowers while taking nectar with its long tongue. Look for the chestnut belt on the abdomen. The moth is often mistaken for a bumble-bee. Winter is passed as a pupa.

SIZE 20–24 mm.

HABITAT Woodland edges and clearings.

FOODPLANT Bedstraws and honeysuckle; also snowberry.

caterpillar

RANGE Much of Europe, except Scotland, Ireland and far N; generally uncommon.

FLIGHT April–September, in one or two broods.

SIMILAR SPECIES Narrow-bordered Bee Hawkmoth (opposite) has narrower borders and black abdominal belt.

Narrow-Bordered Bee Hawkmoth

A day-flying moth which, like the Broad-bordered Bee Hawkmoth, has almost transparent wings and is often mistaken for a bumble-bee, although bumble-bees fly more slowly and do not hover. Look for the narrow wing margins and the black abdominal belt. Winter is passed as a pupa.

SIZE 18–20 mm.

HABITAT Woodland edges and clearings, moors and damp grassland.

FOODPLANT Devil's-bit scabious and field scabious.

RANGE All Europe, including the Arctic; nowhere common and generally declining.

FLIGHT April–July.

SIMILAR SPECIES Broad-bordered Bee Hawkmoth (opposite) has broader borders and brown abdominal belt.

caterpillar

A restless, day-flying moth that makes a clearly audible hum as it darts from flower to flower and hovers in front of them to feed. The forewing is brown and the hindwing is orange, but the moth is usually seen just as a brownish blur. A great migrant, it spreads to most parts of Europe during the summer, sometimes reaching Iceland. The adult overwinters, but rarely survives north of the Alps.

SIZE 21–25 mm.

HABITAT Anywhere with flowers; common in gardens.

FOODPLANT Bedstraws.

RANGE Resident in S; summer visitor elsewhere.

FLIGHT All year in S, in two or three broods.

SIMILAR SPECIES None.

caterpillar

The forewing ground colour may be green, brown, or grey, but the central band is always darker than the rest. The hindwing is yellow with a brown border. The moth often flies by day and is very variable in size. Winter is passed as a pupa.

SIZE 15–30 mm.

HABITAT Woodland clearings and other sunny spots, often near water.

FOODPLANT Willowherbs, evening primrose and purple loosestrife.

RANGE S & C Europe; declining in many areas.

FLIGHT May–July.

SIMILAR SPECIES Lime Hawkmoth (p.180) is much larger.

Rosebay Willowherb

The forewings, which are swept right back into an arrowhead shape at rest, often have a pinkish tinge when fresh. Look for the narrow and often broken brown border at the front. The outer region of the hindwing is pink. Winter is passed as a pupa.

SIZE 28–35 mm.

HABITAT Almost anywhere with plenty of flowers.

FOODPLANT Various spurges.

caterpillar

RANGE Resident in S & C Europe; migrates northwards each spring and summer, but only rarely reaches British Isles and N.

FLIGHT May–September, in two broods.

SIMILAR SPECIES Bedstraw Hawkmoth (opposite) has broader brown border to forewing and pale brown or buff edge to hindwing.

The front edge of the forewing has a fairly broad, unbroken brown margin, usually darker than that of the Spurge Hawkmoth. The outer edge of the forewing is also rather dark. The outer area of the hindwing, beyond the black stripe, is pale brown. Winter is passed as a pupa.

SIZE 28–35 mm.

HABITAT Anywhere with plenty of flowers.

FOODPLANT Bedstraws and willowherbs.

caterpillar

RANGE Resident in S & C Europe; migrates northwards each spring and summer. Rare visitor to British Isles.

FLIGHT May–September, in two broods.

SIMILAR SPECIES Spurge Hawkmoth (opposite). Mediterranean Hawkmoth is bigger, with pale front border.

Elephant Hawkmoth
Deilephila elpenor

This beautiful moth takes its name from its caterpillar, which has a trunk-like snout. The deep bronzy or olive green of the forewing may fade to a yellowish green with age. The hindwing is black at the base, but otherwise deep pink. Look for the moth hovering at honeysuckle and other flowers at dusk. Winter is passed as a pupa.

SIZE 26–32 mm.

HABITAT Light woodland, river-banks, gardens and wasteland.

FOODPLANT Willowherbs, bedstraws and fuchsias.

RANGE Most of Europe except far N.

FLIGHT May–July.

SIMILAR SPECIES Small Elephant Hawkmoth (opposite) is smaller and much yellower.

caterpillar

This picture of a freshly emerged moth drying its wings reveals the yellow hindwing and its pink outer margin. When properly at rest, the wings are laid flat over the body and swept back like an arrowhead. The forewing is mainly pink, with a broad yellow band in the outer half. Look for the moth feeding at rhododendrons and honeysuckle at dusk. Winter is passed as a pupa.

SIZE 20–25 mm.

HABITAT Woodland clearings, heaths and grassland.

FOODPLANT Bedstraws and occasionally willowherbs.

RANGE Most of Europe except far N.

FLIGHT May–July.

SIMILAR SPECIES Elephant Hawkmoth (opposite) is larger and pink and green with black base to hindwing.

caterpillar

The silvery-grey forewings are wrapped around the body at rest. Their rounded buff tips, together with the buff thorax, give the moth an amazing similarity to a broken twig. The gregarious yellow and black caterpillars can completely strip small trees, causing serious damage in orchards and forestry plantations. Winter is passed as a pupa.

SIZE 28–36 mm.

HABITAT Woods, parks, gardens and hedgerows; often common in towns.

FOODPLANT A wide range of deciduous trees.

RANGE All Europe except far N.

FLIGHT May–August.

SIMILAR SPECIES None.

caterpillar

A very hairy moth, better known for the defensive displays of its caterpillar than its adult appearance. The female, shown here, is always darker and more heavily marked than the male, with a fairly prominent grey band crossing the basal area of the forewing. Winter is passed as a pupa, concealed in a very tough cocoon usually attached to tree-trunks.

SIZE 26–36 mm.

HABITAT Woods, parks, gardens; almost anywhere with trees.

FOODPLANT Poplars and sallows.

RANGE All Europe.

FLIGHT April–July.

SIMILAR SPECIES *Cerura erminea* of S & C Europe has prominent black loops in basal area of forewing.

caterpillar

Lobster Moth

Stauropus fagi

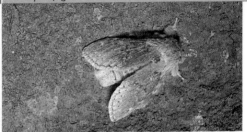

This rather hairy moth is named for its very unusual, long-legged caterpillar. The typical form, shown here, is pale brown, heavily dusted with grey to give it an ashy appearance, but there is also a dark brown form. The moth rests with its hindwings protruding beyond the front edge of the forewings, looking just like a piece of bark. Winter is passed as a pupa.

SIZE 25–34 mm.

HABITAT Woodland.

FOODPLANT Beech, birch, oak and various other trees.

RANGE Most of Europe, except Scotland and far N.

FLIGHT May–August.

SIMILAR SPECIES Great Prominent lacks the zigzag white cross lines and has almost white hindwings.

caterpillar

This moth is named for the colour of the forewing, complete with 'rusty' patches in most specimens. Look for the tuft of scales on the rear of the forewing, standing upright when the wings are folded at rest. Winter is passed as a caterpillar.

SIZE 18–23 mm.

HABITAT Light woodland, heaths, hedgerows and gardens.

FOODPLANT Birch and, less often, alder and hazel.

RANGE N & C Europe.

FLIGHT April–September, in one or two broods.

SIMILAR SPECIES Dark form of Lobster Moth (opposite) is larger, and lacks the 'prominence' on the forewing.

caterpillar

The ground colour of the forewing ranges from fawn or greyish brown to a deep reddish brown. The pattern also varies a little, but the dark, pebble-like patch that gives the moth its name is always visible at the tip. The tuft of scales is always prominent on the rear of the forewing. The hindwing is dirty white. Winter is passed as a pupa.

SIZE 18–24 mm.

HABITAT Open woodland and hedgerows, especially in damp areas.

FOODPLANT Willows, sallows and poplars.

RANGE All Europe.

FLIGHT May–October in 1–3 broods.

SIMILAR SPECIES Three-humped Prominent has much darker forewings and very white hindwings.

caterpillar

The ground colour of the forewing is white, but in many specimens it is heavily clouded with brown. Look for the prominent white wedge-shaped mark near the rear corner of the wing. The hair on the thorax ranges from grey to light or dark brown. Winter is passed as a pupa.

SIZE 20–26 mm.

HABITAT Wooded areas, scrub and heathland.

FOODPLANT Birch.

RANGE All Europe.

FLIGHT April–September, in one or two broods.

SIMILAR SPECIES Swallow Prominent lacks the white wedge, although it has several white streaks.

caterpillar

The ground colour ranges from pale brown to deep chestnut, but the moth can usually be recognized by the distinctive pale 'crown' on the thorax. The tuft of scales on the rear of the forewing is very prominent at rest, with the wings held steeply over the body. Winter is passed as a pupa.

SIZE 24 mm.

HABITAT Woodlands and hedgerows.

FOODPLANT A wide range of deciduous trees and shrubs, especially birch and hazel.

RANGE All Europe.

FLIGHT
May–September, in two broods.

SIMILAR SPECIES Maple Prominent (opposite) has no pale 'crown' and usually has pale wing-tips.

caterpillar

The dark rear margin and the pale outer part of the forewing combine to give the resting moth a passable resemblance to a large bird dropping, although the pale area is not always well developed. Look also for the prominent tuft of scales on the rear of the forewing and the conspicuous brown 'crown' or crest on the thorax. Winter is passed as a pupa.

SIZE 18 mm.

HABITAT Hedgerows and woodland edges, mainly on lime-rich soils.

FOODPLANT Field maple and possibly sycamore.

RANGE C Europe, except Scotland and Ireland.

FLIGHT May–July.

SIMILAR SPECIES
Coxscomb Prominent (opposite) has white thoracic 'crown' and no pale wing-tips.

caterpillar

The chocolate-brown patch near the tip of the forewing is quite bright and conspicuous in moths from Scotland and other northern areas, but often rather subdued in specimens from more southerly regions, as shown here. The thorax and the tip of the abdomen also have dark brown scales. Winter is passed as a pupa.

SIZE 13 mm.

HABITAT Marshes and damp woodland.

FOODPLANT Sallows and willows, especially creeping willow.

RANGE N & C Europe.

FLIGHT April–October, in 1–3 broods.

SIMILAR SPECIES Chocolate-tip is larger and browner, with larger and more obvious chocolate wing-tip.

Grey Sallow

Only the male has wings. The female is little more than a bag of eggs and rarely moves from the surface of her cocoon, where she mates and lays her eggs. The male flies mainly by day and is often seen darting rapidly along hedgerows in search of females. Winter is passed as an egg.

SIZE 16 mm.

HABITAT Almost anywhere with trees and shrubs; common in towns; often a pest of ornamental trees.

FOODPLANT Almost any deciduous tree or shrub.

RANGE All Europe, including Iceland.

FLIGHT June–October, in 1–3 broods.

SIMILAR SPECIES Scarce Vapourer is darker, with pale spots at tip of forewing. *Orgyia ericae* of N & C Europe is paler, with smaller white spot.

Pale Tussock

Calliteara pudibunda

The female, shown here, is larger and paler than the male, which also has a contrasting central band on the forewing. Notice the resting position, with the hairy front legs well forward, shared by several other members of the family. Winter is passed as a pupa.

SIZE 22–32 mm.

HABITAT Woods, parks, orchards and gardens; once common in hop fields, where the hairy larvae gave hop-pickers painful rashes.

FOODPLANT Hops and a wide range of deciduous trees and shrubs.

RANGE N & C Europe, except Scotland and far N.

FLIGHT April–July.

SIMILAR SPECIES Dark Tussock has darker body hair.

♀

Named for the tuft of yellow hairs at the tip of the abdomen. Like several other members of the family, the female covers her eggs with these hairs. The male is smaller and has a smaller tuft than the female, shown here. There is often a dark brown spot at the rear corner of the forewing. Winter is passed as a small caterpillar.

SIZE 15–20 mm.

HABITAT Woods and hedgerows.

FOODPLANT Hawthorn and many other deciduous trees and shrubs.

RANGE Most of Europe except far N.

FLIGHT June–August.

SIMILAR SPECIES
Brown-tail has brown tuft.
White Satin has less hairy legs,
banded with black and white.

caterpillar

Although the female, shown here, has wings, she cannot fly. She stays close to her cocoon, often on a tree-trunk, and waits for the day-flying male. After mating, she covers her eggs with a mass of pale brown hairs from her abdomen. The male is smaller and dark brown with the same pattern of lines as the female and chequered margins. Winter is passed as an egg.

SIZE 22 mm (male) – 30 mm (female).

HABITAT Woodland; a serious forest pest.

♂

FOODPLANT A wide range of deciduous trees and shrubs.

RANGE Most of Europe except far N; extinct in Britain since the early 20th century.

FLIGHT July–September.

SIMILAR SPECIES
Pale Brindled Beauty is like male, but greyer with no chequered margins and flies in spring.

The female is larger than the male and reluctant to fly, but the sexes are otherwise alike. Look for the pink-striped abdomen. Winter is passed as an egg.

SIZE 20–25 mm.

HABITAT Woodland, both deciduous and coniferous.

FOODPLANT Mainly oak, but also birch and various conifers; a serious pest of spruce and pine in some parts of Europe.

RANGE Most of Europe, except Scotland and Ireland; rare in far N.

FLIGHT July–September.

SIMILAR SPECIES None.

caterpillar

Syntomis phegea

This is a day-flying moth with a rather weak, drifting flight rather like some of the burnets. It is sometimes called the Yellow-belted Burnet, although it is not related to the burnet moths. Winter is passed as a caterpillar, in a communal web under rocks and stones.

SIZE 12 mm.

HABITAT Grassland and scrub in sunny spots with plenty of flowers; often in gardens.

FOODPLANT Dandelions and many other herbaceous plants.

RANGE S & C Europe except British Isles.

FLIGHT May–August.

caterpillar

SIMILAR SPECIES
Several in S Europe, differing mainly in spot pattern.

The bright pink of the forewing often fades a little with age, but the moth is always easily recognized. It is often very common in light woodland and scrubby habitats and comes to light traps in large numbers but, like most other footmen and other lichen-feeding species, it is absent from regions with much air pollution. Winter is passed as a caterpillar.

SIZE 10–15 mm.

HABITAT Open woods and scrub.

FOODPLANT Various lichens on tree-trunks and branches.

RANGE Most of Europe, except Scotland and far N and S.

FLIGHT June–August.

SIMILAR SPECIES None.

All four wings are sooty black and the front of the thorax is bright red or occasionally orange. The moth flies by day as well as at night and sometimes swarms around the tops of woodland trees. Winter is passed as a pupa.

SIZE 14 mm.

HABITAT Light woodland and surrounding scrub with plenty of lichens on trees.

FOODPLANT Various lichens.

RANGE N & C Europe except Scotland; rare in far N.

FLIGHT May–July.

SIMILAR SPECIES None.

The front edge of the steely-grey forewing is almost straight and the pale streak at the front tapers and does not quite reach the wing-tip. The wings are laid almost flat over the body at rest. Adults are common visitors to light. Winter is passed as a caterpillar.

SIZE 15 mm.

HABITAT Woods, hedgerows and orchards with plenty of lichens.

FOODPLANT Various lichens.

RANGE Most of Europe except far N.

FLIGHT June–August.

SIMILAR SPECIES
Dingy Footman has strongly curved forewing and pale marginal streak does not taper. Scarce Footman wraps wings tightly round body at rest and marginal streak does not taper.

caterpillar

Lithosia quadra

The male, shown here in its typical resting position, has greyish brown forewings with a distinct golden patch at the base, but the species is named for the much larger female, which has golden yellow forewings each with two black spots. The thorax is golden in both sexes and the hindwing is pale yellow. Winter is passed as a small caterpillar.

SIZE 15 mm (male) – 25 mm (female).

HABITAT Woodland.

FOODPLANT Lichens growing on trees, especially oaks.

RANGE Most of Europe except far N.

FLIGHT July–September, flying by day and night.

SIMILAR SPECIES None.

The patterns of both forewing and hindwing vary enormously. Sometimes the forewing is almost completely brown and the hindwing may be yellow instead of orange, but the species is usually quite easily recognized. Look for the very hairy brown thorax and narrow red collar. Winter is passed as a small caterpillar.

SIZE 25–35 mm.

HABITAT Gardens, wasteland and most open habitats.

FOODPLANT A wide range of low-growing plants.

RANGE All Europe.

FLIGHT June–August.

caterpillar

SIMILAR SPECIES Cream-spot Tiger (p.212) has black thorax and forewing. Jersey Tiger (p.217) is more obviously striped and less hairy.

Cream-Spot Tiger

Arctia villica

The forewing is basically black with cream or white spots, but the pattern varies and the spots may run together. Look for the black thorax with white sides and the pale spot at the base of each forewing. Winter is passed as a caterpillar.

SIZE 22–30 mm.

HABITAT Light woodland, scrub and grassy places; mainly coastal in Britain.

FOODPLANT A wide range of low-growing plants.

RANGE Most of Europe N to S Scandinavia, but not Scotland and Ireland.

FLIGHT May–July.

SIMILAR SPECIES Garden Tiger (p.211) has brown thorax. Scarlet Tiger (p.218) has red hindwings.

caterpillar

The male, shown here, flies by day and night and is easily recognized. Look for the red fringes on all wings. The female is much smaller, with a dark orange forewing and black and orange hindwing. She flies mainly at dusk and dawn. Winter is passed as a small caterpillar.

SIZE 15 mm (female) – 23 mm (male).

HABITAT Heaths, moors, rough grassland and scrub.

FOODPLANT Heathers and many other low-growing plants.

RANGE Most of Europe except far N.

FLIGHT June–August.

SIMILAR SPECIES None.

♀

White Ermine

Spilosoma lubricipeda

The black spots vary a lot in size and density and are sometimes absent, but there is always a spot in the middle of the hindwing. The ground colour is occasionally pale yellow, especially in the north. Look for the yellow abdomen with a white tip. This very common moth often comes to lights in large numbers. Winter is passed as a pupa.

SIZE 15–23 mm.

HABITAT Almost anywhere, including towns and gardens.

FOODPLANT A wide range of herbaceous plants.

RANGE All Europe except far N.

FLIGHT May–September, in one or sometimes two broods.

SIMILAR SPECIES Female Muslin Ermine has white abdomen. Water Ermine is almost or quite spotless and never has spots on hindwing.

caterpillar

The ground colour ranges from very pale cream to deep yellow. The amount of black spotting also varies, but the faint diagonal line is usually visible and so is the large spot on the rear edge of the forewing. The abdomen is entirely yellow with black spots. Winter is passed as a pupa.

SIZE 15–20 mm.

HABITAT Almost anywhere, including parks and gardens.

FOODPLANT A wide range of wild and cultivated plants.

RANGE All Europe except far N.

FLIGHT May–August.

SIMILAR SPECIES Yellowish forms of White Ermine (opposite) lack diagonal line and have white abdominal tip.

caterpillar

The forewing ranges from rich chestnut brown to dull greyish brown. The hindwing is typically rose-pink with dark smudges in the outer region, but in specimens with greyish forewings – found mainly in the north – it is grey with a small pink area at the base. A very common visitor to lights. Winter is passed as a fully-fed caterpillar.

SIZE 12–18 mm.

HABITAT Hedgerows, wasteland and other grassy places.

FOODPLANT Dandelion, dock and many other low-growing plants.

RANGE All Europe.

FLIGHT May–September, in one or two broods.

SIMILAR SPECIES None.

caterpillar

The moth flies by day as well at night and its bright red or orange hindwing often causes it to be mistaken for a butterfly. The forewing pattern varies a little, but the rear edge always has a pale border. Look also for the brown and white striped thorax. Winter is passed as a small caterpillar.

SIZE 28 mm.

HABITAT Light woodland and scrubby places, including gardens and orchards; prefers dry places.

FOODPLANT Dandelion, plantains and other low-growing plants.

RANGE S & C Europe, but only in Devon in Britain.

FLIGHT July–September.

SIMILAR SPECIES Garden Tiger (p.211) looks similar, but is stouter with entirely brown thorax.

caterpillar

Scarlet Tiger

Callimorpha dominula

The black forewing has a metallic green tinge and a variable pattern of white and yellowish spots. The hindwing is scarlet or occasionally yellow, with black spots. Look for the two narrow orange or cream stripes on the thorax. The moth flies by day and at night. Winter is passed as a small caterpillar.

SIZE 26 mm.

HABITAT River-banks and other damp places.

FOODPLANT A wide range of low-growing plants; also sallows and bramble

RANGE Most of Europe, but confined to the southern half of Britain – mainly in the west.

FLIGHT June–August.

SIMILAR SPECIES None.

caterpillar

An unmistakable moth, although the size of the red spots may vary.
The hindwing is scarlet with a narrow black border. The wings are
feeble and flight is weak. The moth is basically nocturnal, but often
flutters away when disturbed in the daytime. Winter is passed as a
pupa. The black and gold caterpillars have been used to control
ragwort in some places.

SIZE 18 mm.

HABITAT Open grassy places, including agricultural set-aside, but
absent from heavy soils.

FOODPLANT Ragwort.

RANGE All Europe, including
Iceland, but not the Arctic.

FLIGHT May–August.

SIMILAR SPECIES None.

caterpillar

Agrotis exclamationis

The moth is named for the prominent dark marks on the forewing, which ranges from straw-coloured to chestnut or greyish brown. The hindwing is pearly white in the male, often with a narrow brown border, and rather cloudy in the female. The wings are laid flat over the body at rest, with a good overlap. Look for the dark collar. Winter is passed as a fully-grown caterpillar.

SIZE 15–20 mm.

HABITAT Almost anywhere; especially common in gardens and other cultivated areas.

FOODPLANT A wide range of herbaceous plants.

RANGE All Europe except far N.

FLIGHT May–September, in one or two broods.

SIMILAR SPECIES Heart and Club has shorter, broader 'dart' and no dark collar.

caterpillar

The forewing ranges from very dark to mid-brown in the male, shown here, but is much paler and often mottled with grey in the female. The hindwing, usually seen only when the moth is disturbed and takes off on its fast, zig-zag flight, is bright yellow with a dark border. Its wings are laid flat over the body at rest. Winter is passed as a caterpillar.

SIZE 25 mm.

HABITAT Almost anywhere; one of our commonest moths.

FOODPLANT Almost any deciduous tree or shrub.

RANGE All Europe, including Iceland.

FLIGHT June–October.

SIMILAR SPECIES Lunar Yellow Underwing has large black spot near centre of hindwing.

caterpillar

Broad-Bordered Yellow Underwing

Noctua fimbriata

The female, shown here, is paler than the male, whose forewing ranges from reddish brown to almost black. Both sexes may have a greenish tinge. The hindwing is deep yellow with a black border almost half the width of each wing. At rest, the wings are held slightly roofwise, with less overlap than in the Large Yellow Underwing (p.221). Winter is passed as a caterpillar.

SIZE 26 mm.

HABITAT Light woodland, parks and scrub.

FOODPLANT Birch, sallow and other deciduous trees and shrubs.

RANGE All Europe except far N.

FLIGHT June–September.

SIMILAR SPECIES
Several other yellow underwings have quite broad borders, but all are smaller moths.

caterpillar

The forewing ranges from greyish to reddish brown, but the pale patch near the front edge is always conspicuous and contrasts sharply with the bow-tie-shaped black patch below it. The hindwing is white with a narrow brown border. The wings are laid flat over the body at rest. Winter is passed as a caterpillar.

SIZE 15–20 mm.

HABITAT Almost anywhere, but mainly in lowlands.

FOODPLANT A very wide range of low-growing plants.

RANGE All Europe except far N.

FLIGHT May–October, in two or more broods; most common in autumn.

SIMILAR SPECIES Double Square-spot (p.224) and Hebrew Character (p.230) lack prominent pale patch and have cloudy hindwings.

caterpillar

Double Square-Spot

Xestia triangulum

The ground colour of the forewing is greyish brown, often with a pinkish tinge, and the wings are laid flat over the body at rest. Look for the two almost square dark spots near the front edge of the forewing. The hindwing is greyish brown. Winter is passed as a small caterpillar.

SIZE 15–20 mm.

HABITAT Woods and hedgerows.

FOODPLANT Bramble and many other trees and shrubs; also docks.

RANGE All Europe except far N.

FLIGHT May–July.

SIMILAR SPECIES
Setaceous Hebrew Character (p.223) has darker forewings, with pale patch on front edge, and white hindwings.

caterpillar

Easily recognized by the conspicuous white spot on the blue-black forewing, although on rare occasions the spot is obscured by the general wing colour. The wings form a shallow roof over the body at rest, with little overlap. Winter is passed as a pupa.

SIZE 16–22 mm.

HABITAT Almost anywhere; especially common in and around human settlements.

FOODPLANT A wide range of herbaceous plants and shrubs, including cultivated species.

RANGE All Europe; rare in far N.

FLIGHT June–September.

SIMILAR SPECIES Cabbage Moth is greyer with much less obvious white spot.

caterpillar

Bright-Line Brown-Eye

Laccanobia oleracea

Look for the bright sub-marginal line, with two prominent spikes near the centre, and the orange-brown, kidney-shaped spot near the middle of the forewing. The ground colour ranges from pale to dark brown. The wings are held roofwise at rest.
Winter is passed as a pupa.

caterpillar

SIZE 15–20 mm.

HABITAT Grassland, marshes (including saltmarshes), wasteland and gardens.

FOODPLANT A wide range of herbaceous plants, especially the goosefoot family.

RANGE All Europe except far N.

FLIGHT May–September, in one or two broods.

SIMILAR SPECIES Broom Moth (opposite) lacks brown 'eye'.

The ground colour of the forewing ranges from greyish brown to purplish brown. The jagged, pale sub-marginal line is often indistinct, but there is always a pale blotch near the rear corner. The wings are held roofwise at rest. Winter is passed as a pupa.

SIZE 15–20 mm.

HABITAT Heaths, moors, grassland and other open country.

FOODPLANT Many trees, shrubs and herbaceous plants, including bracken.

RANGE All Europe, including Iceland.

FLIGHT May–August.

SIMILAR SPECIES Bright-line Brown-eye (opposite) has whiter sub-marginal line.

caterpillar

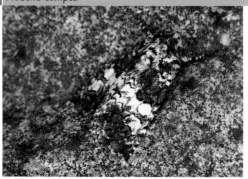

The density of the forewing ground colour varies, but the broad white band running right across the wing separates this moth from several related species. The species was hardly known in Britain until 1948, but has since spread all over eastern England as far north as the Wash. Winter is passed as a pupa.

SIZE 15 mm.

Red Campion

HABITAT Open country on the Continent; almost entirely restricted to gardens in England.

FOODPLANT Developing seeds of pinks and campions; restricted to sweet williams in England.

RANGE Most of Europe except far N.

FLIGHT May–September.

SIMILAR SPECIES Marbled Coronet has broken white band and white blotch at wing-tip.

This moth is named for the pale, branched streak running through the forewing, although this is sometimes reduced to a simple line. The ground colour ranges from pale greyish brown to rich chestnut. The sexes are alike, but the female is very much larger than the male. Winter is passed as eggs, which are broadcast over the grassland.

SIZE 12 mm (male) – 18 mm (female).

HABITAT Rough grassland, mainly in uplands.

FOODPLANT Grasses and rushes, often destroying large areas of upland pasture.

RANGE All Europe, including Iceland.

FLIGHT June–September.

SIMILAR SPECIES None.

caterpillar

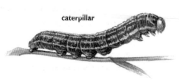

Hebrew Character

Orthosia gothica

The ground colour ranges from grey to reddish brown, but the black mark shaped vaguely like a bow-tie is usually clearly visible. The darker, redder forms occur mainly in northern areas. The hindwing is dingy grey. The wings are held roofwise over the body at rest. Winter is passed as a pupa.

SIZE 16 mm.

HABITAT Almost anywhere.

FOODPLANT A wide range of deciduous trees and shrubs; also many herbaceous plants.

RANGE All Europe.

caterpillar

FLIGHT February–May.

SIMILAR SPECIES Setaceous Hebrew Character (p.223) has conspicuous pale patch at front of forewing and lays wings flat at rest.

Look for the shining white spot in the centre of the forewing and the
two narrow brown cross lines, the inner one of which bends sharply
in the middle to form a right angle. The ground colour ranges from
pale brown to deep orange. Winter is passed as a caterpillar.

SIZE 17 mm.

HABITAT Hedgerows and woodland edges.

FOODPLANT Coarse grasses, especially couch grass and
cocksfoot.

RANGE All Europe except far N.

FLIGHT June–September.

SIMILAR SPECIES None.

The narrow forewings, with dark brown streaks on a pale brown background, are pulled tightly back along the body at rest, making the moth look just like a piece of bark. Winter is passed as a pupa. The species is best known for its conspicuous black and yellow spotted caterpillar, which ruins garden mulleins.

SIZE 20–26 mm.

HABITAT Gardens, scrub and wasteland.

caterpillar

FOODPLANT Mulleins and figworts; sometimes garden buddleia.

RANGE S & C Europe, including S Sweden, but not Scotland and Ireland.

FLIGHT April–June.

SIMILAR SPECIES Striped Lychnis and Water Betony are almost identical, but unlikely to be found in Britain.

The forewing often lacks the green tinge and may be dull chocolate brown with darker markings, but there is always a white fleck close to the rear edge. Look also for the conspicuously chequered margins. The moth is often common at ivy blossom. Winter is passed as an egg.

SIZE 20 mm.

HABITAT Woodland edges, hedgerows and scrub.

FOODPLANT Hawthorn and blackthorn.

RANGE All Europe except far N.

FLIGHT August–October.

SIMILAR SPECIES None.

caterpillar

The delicate green ground colour of the forewing soon fades to yellowish or even to white in older moths and dead specimens. In some specimens the ground colour is yellowish green and the black markings are much lighter than in the moth shown here. Winter is passed as an egg.

SIZE 22 mm.

HABITAT Oak woods and parkland.

FOODPLANT Oak; feeds on buds at first, then moves to flowers and leaves.

RANGE Most of Europe except far N.

FLIGHT August–October.

SIMILAR SPECIES Scarce Merveille du Jour has strong black cross line near base of forewing and flies in midsummer.

caterpillar

The forewing ranges from greyish or straw-coloured to brick-red and the darker mottling is sometimes absent, but the small black dots on the front edge of the forewing will normally identify this common species. Look also for the short oblique streak near the wing-tip. The moth is often abundant at ivy blossom. Winter is passed as an egg.

SIZE 18 mm.

HABITAT Hedgerows and scrubby places.

FOODPLANT A wide range of herbaceous plants when young; often moves to shrubs later.

RANGE S & C Europe; rare in Scotland and Ireland.

FLIGHT September–November.

SIMILAR SPECIES The Brick has no streak near wing-tip and less obvious spots on front edge.

caterpillar

The ground colour of the forewing is usually pale yellow, but is sometimes bright yellow or even orange. The darker areas range from deep pink to purplish brown. The moths frequently feeds at ivy blossom. Winter is passed as an egg.

SIZE 16 mm.

HABITAT Beechwoods and surrounding hedgerows and scrub.

FOODPLANT Beech and field maple; feeds on buds at first, and later on flowers and leaves.

RANGE Most of Europe except far N.

FLIGHT August–October.

SIMILAR SPECIES The Sallow (opposite) and Pink-barred Sallow have outer part of forewing yellow.

caterpillar

The ground colour of the forewing ranges from pale yellow to orange, and sometimes the wings are almost entirely yellow, with no dark markings. The hindwing is pure white. The moths commonly gather to feed on over-ripe fruit. Winter is passed as an egg.

SIZE 16 mm.

HABITAT Hedgerows, woodland edges and scrub.

FOODPLANT Sallow catkins at first, then various herbaceous plants when catkins fall.

RANGE All Europe.

FLIGHT August–October.

SIMILAR SPECIES Barred Sallow (opposite) has purplish outer areas. Pink-barred Sallow has purplish collar.

caterpillar

Grey Dagger

Acronicta psi

Named for the small black dagger-like markings on its pale grey forewing, although melanic forms with very dark wings occur in many areas, especially in industrial regions. Winter is passed as a pupa. The Dark Dagger is virtually identical and can be reliably distinguished only by examining the reproductive organs, although the caterpillars are very different.

SIZE 15–20 mm.

HABITAT Woods, parks, gardens and hedgerows.

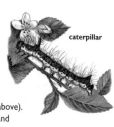

caterpillar

FOODPLANT Hawthorn, blackthorn and many other deciduous trees.

RANGE All Europe except far N.

FLIGHT June–July

SIMILAR SPECIES Dark Dagger (see above). Miller Moth is much paler, with fewer and smaller black marks.

Named for its coppery-orange hindwings, this moth rests with its wings almost flat. Look for the clearly chequered sides of the abdomen, especially noticeable on the underside. The underside of the hindwing is largely straw-coloured with an orange tinge near the outer edge. Winter is passed as an egg.

SIZE 25 mm.

HABITAT Open woods, parks and hedgerows.

FOODPLANT Deciduous trees and shrubs, especially oak.

RANGE All Europe except far N.

FLIGHT June–October.

SIMILAR SPECIES
Svensson's Copper Underwing is very similar, but abdomen has no obvious chequering and underside of hindwing is largely grey.

caterpillar

Old Lady

Mormo maura

This rather drab moth rests with its wings laid almost flat, although there is hardly any overlap. It is often found roosting in sheds and other buildings in the daytime. The pattern on its forewing is reminiscent of the shawls once worn by elderly ladies. The hindwing is dull brown. Winter is passed as a small caterpillar.

SIZE 35 mm.

HABITAT Light woodland, hedgerows and gardens.

FOODPLANT Herbaceous plants at first; later on blackthorn and many other trees and shrubs.

caterpillar

RANGE S & C Europe; rare in Scotland and Ireland.

FLIGHT June–August.

SIMILAR SPECIES Red Underwing (p.251) is similar at rest, but easily distinguished by its red hindwings.

This very common moth habitually rests with the front edge of the forewing rolled or folded in a very unusual position, giving it a remarkable similarity to a crumpled leaf. The ground colour ranges from olive green to a rosy pink, and the V-shaped band ranges from green to chestnut. Winter is passed as a caterpillar.

SIZE 20–25 mm.

HABITAT Almost anywhere.

FOODPLANT A wide variety of herbs, trees and shrubs.

RANGE Most of Europe, but only as summer visitor in N areas; commonly reaches Iceland.

FLIGHT All year, but mainly May–October, in one or two broods.

SIMILAR SPECIES None.

caterpillar

The male, shown here, has red fringes and a yellowish hindwing. The female has paler fringes, with more white dusting on the forewing and a pure white hindwing. The silvery lines are bordered with dark green in the British race, shown here. 1st-brood insects have three lines on each forewing, but 2nd-brood moths, rarely seen in Britain, have only two lines. Winter is passed as a pupa.

SIZE 16 mm.

HABITAT Wooded regions and hedgerows.

FOODPLANT Oak, beech and some other deciduous trees.

RANGE Most of Europe except far N.

FLIGHT May–September, in one or two broods.

SIMILAR SPECIES Scarce Silver-lines is larger, with only two silvery lines and pale yellow fringes.

caterpillar

The metallic areas of the forewing range from brassy green to deep gold and there is often a complete brown band across the middle of the wing. The moth rests with its wings held steeply roofwise and, like most of its relatives, it has a prominent thoracic crest. Winter is passed as a small caterpillar.

SIZE 15–20 mm.

HABITAT Almost anywhere; very common in gardens and waste places.

FOODPLANT Stinging nettle and many other herbaceous plants.

RANGE All Europe.

FLIGHT May–October, in one to three broods

caterpillar

SIMILAR SPECIES Slender Burnished Brass has broad brown border at front and rear of forewing.

The silvery figure 8 on a golden background makes this moth very easy to recognize, although the intensity of the golden colour varies from pale yellow to orange-brown. Look for the long palps held up in front of the face at rest. Winter is passed as a small caterpillar.

SIZE 20 mm.

HABITAT Mainly parks and gardens, especially in British Isles and other more northerly areas.

FOODPLANT Monkshood and cultivated delphiniums; often causes serious damage in gardens.

RANGE Most of Europe except far N.

FLIGHT June–September, in one or two broods.

SIMILAR SPECIES None.

Delphinium

The golden ground colour varies in intensity and the darker areas range from light brown to chestnut. The silver (not gold) spots occasionally meet in the middle. Winter is passed as a small caterpillar.

SIZE 18 mm.

HABITAT Damp places, both wooded and open.

FOODPLANT Sedges, yellow iris and other plants of damp places.

RANGE All Europe except far N.

FLIGHT June–October, in one or two broods.

SIMILAR SPECIES Lempke's Gold Spot has more golden appearance and rounder outer silver spot.

Yellow Iris

The ground colour ranges from silvery grey to velvety black, with a purple tinge. Look for the central silver Y and the diagonal black streak at the wing-tip. The moth flies by day, is most often seen as a greyish blur while feeding at flowers, and is abundant in gardens. It overwinters in any stage in the south, where it flies all the year, but rarely survives north of the Alps.

SIZE 15–25 mm.

HABITAT Almost anywhere with flowers.

FOODPLANT A very wide range of herbaceous plants.

RANGE Resident in S Europe; migrates to all parts in summer, including Iceland.

FLIGHT May–October in most areas; two or more broods.

SIMILAR SPECIES Ni Moth is usually smaller, and lacks dark streak at wing-tip.

caterpillar

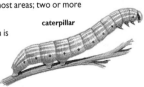

The golden Y is usually broken into two sections and may be reduced to a single spot. It sits on a rather indistinct dark patch sometimes heavily marked with orange. Look for the gold-edged kidney-shaped spot between the Y and the wing-tip. Notice also the prominent abdominal crests typical of this genus. Winter is passed as a caterpillar.

SIZE 20 mm.

HABITAT Hedgerows, gardens and scrubby places.

FOODPLANT Stinging nettles, dead-nettles and many other plants.

RANGE All Europe.

FLIGHT May–August.

SIMILAR SPECIES Plain Golden Y (p.248) has Y on plain brown rectangle, and no gold-edged spot.

Red Dead Nettle

Plain Golden Y

Autographa jota

The golden Y, which is normally broken into two halves, sits on a plain brown rectangle, while the rest of the forewing has a strong pink or purple tinge. Winter is passed as a caterpillar.

SIZE 20 mm.

HABITAT Almost any flowery habitat, including gardens.

FOODPLANT Stinging nettle and a variety of other plants, including trees and shrubs.

RANGE All Europe except far N.

FLIGHT May–August.

SIMILAR SPECIES Beautiful Golden Y (p.247) lacks clear brown rectangle and has gold-edged kidney-shaped spot on forewing.

caterpillar

Named for the two rings of pale hair on the front of the thorax, which look very much like a pair of spectacles, this moth has a purplish grey forewing with silvery streaks at the base and towards the outer edges. The central area is often much darker than the rest of the wing, especially in more southerly areas. Winter is passed as a pupa.

SIZE 15 mm.

HABITAT Open woodland, hedgerows and other rough places.

FOODPLANT Stinging nettle.

RANGE All Europe except far N.

FLIGHT May–September, in one or two broods.

SIMILAR SPECIES Dark Spectacle has darker thoracic rings and lacks silvery streaks on forewing.

caterpillar

Ephesia fulminea

The forewing ranges from pale to dark grey, with conspicuous wiggly lines, and often has a strong violet tinge. The hindwings are deep yellow with a dark brown border, broken at the rear of the wing to create an isolated spot at the rear corner. There is also a brown ring in the basal half of the hindwing. Winter is passed as an egg.

SIZE 28 mm.

HABITAT Woodland edges, hedgerows and orchards.

FOODPLANT Blackthorn and cultivated plums; also oaks.

RANGE S & C Europe.

FLIGHT June–August.

SIMILAR SPECIES
Ephesia nymphaea is browner, without full ring in basal half of hindwing.

Blackthorn

The forewing is essentially grey with darker lines and a variable dusting of brown scales, and when at rest on a tree-trunk the moth is very difficult to detect. When disturbed, it flashes its red and black hindwings, which have conspicuous white fringes. Winter is passed as an egg.

SIZE 38 mm.

HABITAT Light woodland, parks and gardens.

FOODPLANT Willows and poplars.

RANGE All Europe; rare in far N.

FLIGHT July–October.

caterpillar

SIMILAR SPECIES
Dark Crimson Underwing has inner band of hindwing shaped like a W. Light Crimson Underwing has a nearly straight band, and pale patches on forewing.

The jagged wings and the coloration make this moth very easy to identify, although the amount of orange varies and the white cross lines are not always very clear. Winter is passed as an adult, often resting head-down in sheds and other buildings and looking just like a dead leaf.

SIZE 22 mm.

HABITAT Woodland, parks and gardens.

FOODPLANT Willows and poplars.

RANGE Most of Europe except far N, but uncommon in S.

FLIGHT June–November, in one or two broods; again in spring after hibernation.

SIMILAR SPECIES None.

caterpillar